THE GEOGRAPHY OF TOWNS

GEOGRAPHY

Editors:
PROFESSOR S. W. WOOLDRIDGE, c.b.e., d.sc., f.r.s.
and
PROFESSOR W. G. EAST, m.a.
Professors of Geography in the University of London

THE GEOGRAPHY
OF TOWNS

ARTHUR E. SMAILES, M.A.

*Professor of Geography,
University of London*

HUTCHINSON UNIVERSITY LIBRARY

LONDON

HUTCHINSON & CO. (*Publishers*) LTD
178–202 Great Portland Street, London, W.1

London Melbourne Sydney
Auckland Bombay Toronto
Johannesburg New York

First published 1953
Second, revised, impression 1957
Third impression 1958
Fourth impression 1960
Fifth impression 1961
Sixth impression 1962

To
W. L. DIX

This book has been set in Imprint type
face. It has been printed in Great Britain on
Antique Wove paper by Fisher Knight &
Co. Ltd., St. Albans, Herts.

CONTENTS

LIST OF MAPS

CHAPTER I

THE ORIGIN AND BASES OF TOWNS:

I. OLDEN TOWNS

IT is an old saying that cities are as ancient as civilization, yet the etymological kinship expresses a depth of truth that is not always fully appreciated. The rise of civilization has been intimately bound up with the gathering of men to live in cities because the two phenomena have a common geographical basis. From the relations between human societies and the land upon which they are settled city and civilization require the emergence of certain common conditions without which neither can exist. Urban communities can be supported only when the material foundations of life are such as to yield a surplus of food over and above the consuming needs of the food-producers, and when the means are also available to concentrate this surplus at particular spots. Equally, unless society has reached this stage, the opportunities for that accumulation of knowledge and transmission of experience which is civilization remain extremely limited. In the absence of trained specialist classes whose business it is to develop and pass on the social tradition, progress can only be slow and restricted, and must always be precarious.

For most of the time man has existed as a species, the energies of almost the whole of the human race have been claimed in providing the minimum of food, clothes and shelter. Every man who devoted his time to buying and selling, or to any profession, whether military or civil, was a man less to work in getting food, and primitive societies, almost wholly preoccupied in toiling for the bare necessities of life, can afford few such. The professions only become possible as each man working on the land becomes able to grow more food than his family needs. Thus it is no mere accident that in human history the invention of writing and the appearance of city life

7

are twin features that date from the fourth millennium B.C. Cities ever since have been the chief repositories of social tradition, the points of contact between cultures, and the fountainheads of inspiration.

Cities flowered first in the Middle East in the countries that are now Egypt, Iraq and Pakistan. Their appearance was accompanied by great advances in human knowledge and technical equipment. Especially noteworthy among these were a greatly extended use of metals, the invention of the sail, the application of the wheel to transport and more locally to the making of pottery, the invention of the plough and the domestication of animals for draught purposes. Significantly, these were developments which either made possible a great increase in production or facilitated transport. The emergence of urban groups and the new organization of society involved were so closely connected with these other manifestations of a quickened tempo of cultural progress that we are justified in characterizing this phase of human history as an urban revolution. As will be seen later, it was in fact the first of two transformations of comparable significance in the development of human society that warrant this designation.

The scenes of this first urban revolution were areas of irrigation agriculture established in the lower valley of the Nile, in the delta lands at the head of the Persian Gulf, and in the plains of the Indus. In these early homes of settled life, based upon grain cultivation, the farmers, by availing themselves of the regular seasonal river-floods and by developing the use of the plough, were able to raise food-production to a capacity hitherto unknown. For the first time there appeared a sufficient increment to support considerable numbers of people who were not food-producers. Classes were freed to perform for society specialized functions which the newly acquired techniques not only made possible but even demanded for their full application. They gathered in clusters to organize and discharge these special services, and thus there arose towns such as Aphroditopolis and Hierakonpolis in the Nile valley, and Mohenjo-daro and Harappa, 400 miles apart in the plains of the Indus. Early manifestations of urban life, however, were especially notable in the alluvial plains near the

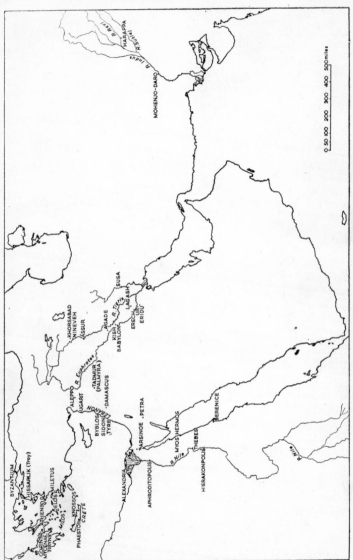

Fig. 1. Some ancient cities of the Middle East

head of the Persian Gulf. Here were Susa in Elem and the cities of Sumer (Ur, Lagash, Erech and others), situated among the former distributaries of the Euphrates. Farther inland, near the site of the later Babylon, was Kish.

When, considerably later it would seem, towns such as Anyang first appeared in China it was with an essentially similar basis, the increment yielded by irrigation agriculture in the fertile riverine lands of the Hwang-ho. And when the archaeological record first discloses civilization established in the New World, the same association is again apparent. Urban life is supported by a surplus provided by developed agriculture. Lacking such advantages for intensive food-production as were possessed by the irrigators of the Old World, the civilization of the Mayas which flourished during the first millennium A.D. depended upon maize cultivation. Its earliest seats were the cities of the Old Empire in Guatemala, such as Copán, Quirigua and Palenque. The abandonment of these cities and the surrounding agricultural areas, with migration northwards into the Yucatan peninsula, where new cities were set up at Mayapán, Uxmal and Chichen, may well have been due to pressure of population overtaxing the soil resources of the earlier lands.

By concentrating the local surplus produced by intensive agriculture the earliest towns were supported. As compared with anything that had gone before, they were settlements distinctive in size, function and appearance. Their inhabitants had created a new type of environment where they pursued a distinctively urban way of life. It was intensely social, highly articulated, and artificial in the sense of being separated from such immediate and constant contact with the soil and the elements. Yet it is important to realize how local was the basis of these earliest urban centres. Each had its surrounding territory upon which it depended for its maintenance. Although the cities, by their very nature, were specialized communities in which division of labour was carried far, many citizens were at least part-time farmers who cultivated the nearby fields. As concentrated groups of settlement the cities were of a size that surpassed anything earlier, but by modern standards their populations must have been small, and their relation with the immediately surrounding countryside was intimate and direct. They were sustained by it. Evidence revealed by excavation of town-sites in Mesopotamia and Pakistan suggests that the population of Sumerian cities may have ranged from 7,000 to 20,000 and that Harappa and Mohenjo-daro approximated to the higher figure. In the absence of thorough excavation of the Egyptian, Chinese and Maya city sites, we can only infer similar figures.

Earlier types of production had been too inefficient to yield a surplus or too diffused for it to be possible effectively to concentrate the individual surpluses of many small producers. Moreover, the first urban groups were restricted in distribution by the localized occurrence of the geographical conditions that made them possible. Before the days of deep ploughing with an ironshod ploughshare it was beyond the capacity of the climate and soils of northern Europe to provide such a surplus. Only the greater yield of crops in a reliable climate of heat and sunshine, by irrigation systems that permitted cropping year after year on areas within reach of the floodwaters and their replenishing silt, enabled the small areas within range of supply to the earliest cities to support concentrations of population as large as they did. A like population density was unattainable elsewhere, as it still is today in those parts of the world where primitive economies of hunting, gathering, pastoralism or shifting cultivation prevail. Furthermore, in the riverine lands of the ancient civilizations physical concentration of the surplus was facilitated by the use of water-transport on the rivers.

The enlarged territorial scope was reflected politically in the efforts made to unify large areas in Egypt and Mesopotamia, where the first empires made their appearance. In the assertion of primacy by successive cities, with subordination of the others and some measure of fusing of their spheres of influence into an empire, may be discerned the beginnings of an urban hierarchy. The empire of Akkad (Agade) which Sargon established after 2752 B.C. and the Babylonian empire of Hammurabi at the close of the third millennium B.C. extended unified organization over the whole of the plains of the Tigris and Euphrates. In consequence the tribute drawn to the political capitals from the much extended area supported considerably larger urban concentrations. Even so, much later than this, in the sixth century B.C., the Babylon of Nebuchadnezzar's empire at the time of the Jewish captivity is estimated to have had a population of not more than eighty thousand. Yet it was probably the most populous city of ancient Mesopotamia. Nineveh, which it superseded, encompassed a larger area (1,400 acres) but was much less built up. We cannot take literally the population figure of more than 120,000 suggested in the Book of Jonah, where Nineveh is described in terms which exaggerate its population as they obviously do its extent. Thebes, biggest capital of the dynastic periods in Egypt, also occupied a large area that was far from being built up. Much semi-rural country was interspersed with the main nuclei, at modern Karnak and Luxor. The figure of 80,000 with which we have credited Babylon far surpasses anything known from archaeological evidence about other cities.

Considerably later than the earliest cities, which were the products of the local increment of concentrated and intensive agricultural activity, trading towns began to appear. They were secondary outgrowths, acquiring their wealth from the services their people rendered as seekers and suppliers to the ancient civilizations of commodities that were either rare luxuries or necessities, such as metals, highly localized in their geographical occurrence. They depended upon the exploitation of the advantages of situations geographically favoured for developing trade relations.

As early as 2000 B.C., before the Bronze Age, Phylakopi, on the island of Milos in the Aegean, had become a centre of the obsidian trade derived from that island, and on the Levant coast Byblos grew up to supply the timber needs of Egypt from its hinterland in the Lebanon. The second Hissarlik, most long-lived of the cities that successively occupied the site later known as Troy, exploited the situation commanding the Dardanelles. In the search for minerals it developed far-reaching trade connections in central Europe. The cities, notably Knossos and Phaistos, that were the seats of the Minoan civilization which flourished in Crete during the first half of the second millennium B.C. likewise derived their wealth

from maritime trade, especially that of Egypt. They were in
turn succeeded in the sixteenth century B.C. by others on the
mainland of Greece, Tiryns and Mycenae. On the Levant
coast, with Byblos as their precursor, the Phoenician cities
Tyre, Sidon and Ugarit grew as trading communities, sustained
by the needs of the imperial powers which successively established themselves in the Levant and whose overlordship the
Phoenicians recognized.

Thus cities appeared on the islands and in the coastlands of
the Mediterranean, where the fragmentation of the cultivated
land set very exacting limits to the size of population group
that could be supported by the local resources. They were
communities which included traders and craftsmen engaged
in supplying the chief markets of the ancient world, exploiting
their intermediate position for that purpose. The trade they
carried on, however, was not a trade in staple foods and they
themselves depended upon local hinterlands for their food
requirements, so that their distribution was clearly related to
pockets of lowland between mountains and sea.

The limitations upon land-transport in ancient times largely
confined the role of trading city to nodes that could be reached
by ships. Caravan cities were a special class, and were not
numerous. The relations of the Phoenician cities with the
populous areas of the ancient Middle East on their landward
side were through the cities of the desert margins, such as
Aleppo and Damascus, and oasis islands such as Tadmur
(Roman Palmyra). These caravan cities were the ports of the
desert, served by its ships the camel caravans. Like the trading
communities in the coastal lowlands of the Mediterranean
their wealth derived from trade, but they originated from and
were sustained by local tracts of fertile cultivated land. Damascus, seventy miles inland from the Mediterranean beyond the
mountains, was the urban centre of the Ghuta, a fertile plain
about 150 square miles in extent, formed by the coalescing
alluvial fans of rivers as they emerged from the mountains.

Port sites abounded in the Mediterranean world of islands
and peninsulas, but urban growth was circumscribed by the
scant resources of the land upon which the trading towns so
largely depended for their food. Specialized agriculture and

bulk trade in grain did not become important until classical times. When the population outgrew its local food supply a new community hived off and established itself across the sea. This happened again and again in the restricted patches of agricultural lowland set amidst the stony wilderness of broken highland which hemmed them in on the landward side and came right to the coast as bounding headlands. From time to time the appearance of military conquerors undoubtedly caused groups of traders from Greek cities, as from those of Crete earlier, to migrate as refugees to new centres where they could pursue their vocations freely. Even so, Greek colonization during its most active period, the eighth and seventh centuries B.C., was agricultural rather than commercial in motive. It arose from the local overpopulation in the limited tracts of agricultural land. The same is probably true of Phoenician colonization and maybe of the Etruscan migration which brought a people from Asia Minor to found cities in the north-west of peninsular Italy about the eighth century B.C. In due course city life was propagated in this fashion from the Aegean into the coast-lands of Asia Minor, westwards into Sicily and southern Italy, North Africa and even into the western basin of the Mediterranean, where Lacydon (Marseilles) was established about 600 B.C. From Miletus alone it is known that more than eighty separate colonies went out.

An urban nucleus was an essential feature of the Greek state. It was contained in but was by no means co-terminous with the city as conceived in classical Greece. Failure to distinguish the two is partly responsible for some of the current misconceptions regarding the size of urban groups in classical times. Such evidence as there is concerning the population of classical Athens points to a citizen population of about 20,000, of whom about three-quarters were peasants living on the land in the lowlands of Attica. These comprised an aggregate area of about one hundred and twenty square miles of agricultural land, divided among four plains. In addition to Athens, the market and centre of political life, a concentration of trade and industry existed four to five miles away at its port, the Piraeus. Even if we allow for the presence of a large number of foreign merchants (*metics*) who are believed to have been nearly half as numerous as the citizens, and of slaves, estimates of whose numbers have been put as high as half the total population, we still have a figure of only 60,000 to 70,000 people for the city state, including rural as well as urban residents. Whatever reasonable allowance we make for these very rough estimates being understatements, others that have assumed the population of classical Athens to have been as great as half a million can only be treated as fantastic. Even the modest figure that has been suggested above was only possible by reliance upon imported supplies of grain, and it was the stranglehold of blockade, cutting off these essential supplies of grain from the Black Sea, that brought Athens to her knees in 405 B.C.

That some of the Greek cities, notably Athens, were able in classical times to outgrow their limited local food-supplies, has often been attributed to a further extension of economic specialization by which the necessary imports were paid for not only by trade services, but by the export of manufactures. Miletus became famous for woollen textiles and Corinth too was renowned for manufactured products. Although Athens' export trade first became important for primary products, oil and wine, arising from agricultural specialization, the products of the Laurion silver-mines and of the workshops of the Piraeus, turning out pottery and metal-goods by slave labour, also became important later. Some authorities, however, believe that the industrial basis of the Greek city state has been exaggerated. They prefer to attribute urban growth in classical Greece to the assertion by the city states of their sea-power, through plunder and systematic exploitation of overseas dependencies. Whether or not the organization of large-scale trade in corn, which was the decisive factor making possible expansion of urban communities in the Mediterranean setting, was essentially a product of political imperialism from its first appearance in the fifth and fourth century Greek city states, it certainly was so in the Hellenistic and Roman empires that followed.

Thanks to the efficiency of its *élite* as soldiers and administrators, Rome became a city on a new scale of size and magnificence. From small beginnings when it depended upon the local agricultural surplus yielded by peasant farmers in the small mountain-girt and hill-studded plains of Latium, the field upon which it drew for its food-supply was extended by political aggrandizement. The Empire became organized to pay tribute to Rome; its surplus production was demanded not only to maintain a large standing army and bureaucracy throughout the conquered territories but also to provide bread and circuses for an idle populace at Rome. The vast size of the area under unified political control, knit together by roads that were extensions of shipping routes, was reflected in the size of the urban concentration which was its nerve centre. There is little trustworthy evidence to enable us to offer a figure for the population of imperial Rome, but as for Athens and Byzantium, some of the high figures that have been suggested are quite improbable, and half a million may be an exaggeration.

In a way that the political organization of classical Greece had not permitted, the empire of Alexander the Great and still more that of Rome opened up opportunities for trade and new fields for extension of city life. Following the conquests of Alexander, cities were founded in the Middle East as colonies of officials, traders and retired soldiers. This process was continued and carried farther afield, especially in Europe to the west and north, by Rome, whose organization of conquered territories was essentially municipal.

Outside Rome itself, the greatest among the trading centres that flourished under the *pax Romana* was Alexandria. Once the foci of civilizations had shifted to the Mediterranean it was natural that an important urban centre should arise in the Nile delta. This development was foreshadowed by Naucratis, a Greek colony of Miletus founded about 650 B.C., but it was Alexander who created the city that bears his name, and in the Roman world Alexandria, served by a great system of waterways, concentrated the corn traffic from Egypt, whose fertile lands became the chief granary of Rome. Although essentially a port and owing its greatness to that fact, Alexandria was much more than a community of traders. As the administrative centre of the province of Egypt and second city of the Empire, urban life manifested itself in many aspects. Alexandria was a great centre of culture and learning, and of manufacture, as well as of trade and administration.

Much more specialized were port-towns such as Ostia at the mouth of the Tiber, where the corn-fleets were received for Rome, Puteoli (near Naples) whither iron-ore was shipped from Elba for manufacture, Carthago Nuova the outlet for the great silver-lead mines of south-east Spain, and Myos Hermos, Arsinoe (Suez) and Berenice, the Red Sea ports of the eastern trade. Most remarkable of all the specialized trading centres, however, was Delos. This satellite of Athens had profited successively from the destruction of both Corinth and Carthage, and after the formation in 132 B.C. of the Roman province of Asia it entered upon its great period as a centre of trade in grain and luxury products, but above all in slaves, from the eastern Mediterranean and Asia. On its tiny island there gathered a highly specialized mart community of 20,000 to 30,000. Under Roman rule, too, the caravan cities of the Syrian desert and its margins, such as Petra and Palmyra, flourished through canalization of trade along particular routes. The way in which these changed from time to time was reflected in the chequered urban fortunes.

Besides essentially trading towns, among which London must be included as a creation of Rome, the Romanization of the Empire was effected through the agency of inland towns and the great highways that connected them. It was an urban culture that was introduced into the conquered lands. At the great nodes of the arterial road system there arose important administrative centres, such as Lugdunum (Lyons). Moreover, the local territorial organization was also focused upon towns. The 'civitas' identified a town with its surrounding area, and the non-citizen population of the countryside was 'attributed' to various urban centres, *coloniae* or *municipia*, according to their origin. The Romans grafted their administrative organization on to the tribal system they found in existence, and the associations between their urban pattern and the tribal territories were close.

The towns administered and in turn depended for their sustenance upon the surrounding tribal areas or *pagi*. Many of the local regional names (*noms de pays*) in which France is so rich derive from this association. As in the Hellenistic empire many urban foundations were colonies of veteran soldiers or Roman citizens from Italy settled overseas as a means of relieving the population problem in peninsular Italy. Other towns, some of which were to have great histories, developed from the civilian settlements (*canabae*) of traders and camp-followers that sprang up beside the great military garrisons and outposts of the empire, e.g. Cologne (Colonia Agrippensis), Belgrade (Singidunum), and York (Eboracum).

In its later phases of decay the Roman Empire became increasingly impoverished under the strain of maintaining its army, bureaucracy, and idle proletariat. Successive tracts of the once closely knit empire lapsed into self-containedness, and as the Empire fell into disorganization the civilized world, and with it the city idea, suffered geographical contraction in northern and western Europe. They lived on together in the eastern Mediterranean, where Byzantium became the new Rome, and at its zenith attained a size comparable with that of imperial Rome. A product of sea-power and largely dependent upon it, Byzantium, thanks to its specialized economy, was able to give as well as to receive. In return for food and raw materials, it distributed the products of its workshops, among which silk textiles were especially famous.

Within the Moslem Empire, too, urban life survived, and even flourished in new administrative and trading centres such as Cordoba and Granada, but in northern Europe the old channels of trade dried up as the old political and ecclesiastical organization shrivelled. In place of an urban civilization, the barbarian societies practised a subsistence economy and rural way of life. No longer was there any incentive to rebuild the towns which had been sacked and pillaged. Insofar as the largely deserted towns were still occupied at all and the concept of the city kept alive through the centuries of urban eclipse in the Dark Ages, this was due to the Church, the only supra-local organization. Its diocesan government retained the old geographical structure whereby a town, the bishop's seat, was linked with a surrounding area of countryside. It was understandable therefore that when the agricultural countryside began to yield a surplus again and trade revived, the focus of this new life in the market-place often grew up alongside the great church. Characteristically, the twin features, church and market-place, dominated the plans of many towns, and mediaeval urbanism is epitomized in the cathedral city.

As Pirenne has emphasized, the reappearance of urban life after the Dark Ages was a product of the revival of trade and growth of population, and the traders who created the mediaeval towns sought security for their activities. The gathering points of the new mercantile communities were places of relative security and protection which in a troubled countryside were usually the defensive sites already occupied by lay or ecclesiastical lords. Except in the lands of contemporary agricultural colonization, few of the towns that sprang up from the tenth century were established on virgin sites. They almost invariably developed from trading communities that attached themselves to some earlier settlements, in the shelter of which they could function and for whose needs they ministered in the first place.

Although it might be an episcopal city, which in turn might occupy the site of a Roman town, the gathering point was most often the fortress of a military lord. The prominence among town-names of elements such as 'burgh' and 'chester' bears witness to this. The 'burgh' was in origin a fortified enclosure, and although it commonly happened that the trading settlement attached to such a pre-urban nucleus developed into a town, it would be a mistake to assume that all 'burghs' became towns. The name cannot by itself be taken as indicative of urban functions. The urban connotation was derived from the common association later, and we must beware of reading back the meaning to a period when 'burgh' had no such significance. In Britain the Anglo-Saxon 'burgh' or 'caester' seems to have been essentially military and agrarian, but from Norman times trading communities gathered, and in course of time enclosed themselves in walled towns alongside the castle-baileys. They also acquired special liberties and privileges, but such franchises, far from being a recognition of existing urban functions, were often granted in anticipation of developments that never materialized. Borough and town are not synonymous terms.

In the lands of mediaeval German colonization east of the Elbe the walled town was a pioneer feature of the German occupation of an alien Slav countryside. The progressive agricultural settlement of the country was accompanied by the planting of planned towns in which castle, church, and market-place expressed the characteristics of the colonizing movement. Earlier, Swedes had planted their trading colonies in the form of palisaded depots (*gorods*) along the great rivers of eastern Europe as they developed trade relations with Byzantium. In this way, the germs of the city idea were introduced into Russia.

The filling in of the rural pattern of northern and western Europe by agricultural colonization accompanying forest clearance and marsh drainage in the rich valley lowlands was the great human achievement of the Dark Ages, whence derived the increment which generated the nascent flow of trade by the tenth century. But under the mediaeval agricultural economy the rural population density was low, and under feudalism the extent of territories over which political organization was effectively unified was small. The areas organized to the support of towns were likewise restricted and few mediaeval towns exceeded 10,000 inhabitants. In England there were only London and York. Most were far less, and only the very largest European cities in the Middle Ages exceeded 50,000. These belonged to two classes. Either they were the capitals of especially rich and powerful units or they were *entrepots* which by virtue of commercial functions and extended trade relations had become less dependent upon their immediate environs. London, with 50,000 inhabitants by 1400, partook of the nature of both classes. Notable among the former was Paris, which grew with each extension of Capetian power in the Paris Basin and beyond. Its population had probably reached a quarter of a million early in the fourteenth century. The commercial and industrial cities of the Low Countries and Rhineland, Bruges, Ghent, Cologne and others, and of northern Italy, such as Venice, Genoa, Florence and Milan, were also metropolitan towns, the nodes of long-distance traffic. Bücher distinguished these from the mediaeval market-towns, locally based in reciprocal relationship with a rural district. The trading population of all towns was augmented by artisans, but the chief concentrations of craftsmen were the major trade centres where raw materials supplied by merchants were worked up and craft gilds grew out of merchant gilds. Later in the Middle Ages, before steam-power brought about a new concentration, manufacturing industry tended to scatter from the towns into the countryside. In migrating, it often left a problem of urban unemployment and stagnation or decline in the sizes of towns. Plague and other factors, some of them obscure, also contributed to reduce the populations of most European towns during the fourteenth and fifteenth centuries.

The metropolitan towns were exceptional under mediaeval conditions. The typical town was nourished by its local agricultural surroundings, its size closely related to the local density of the rural population. But if there were severe limits set to the size of individual towns, save only those with exceptional advantages for assembling a very diffused surplus, the Middle Ages saw a great proliferation of small towns and extension of the area colonized by them. By the end of the Middle Ages the Holy Roman Empire was studded with three thousand corporate towns. Multiplication of market-towns reflected the fragmentation of the mediaeval political and economic structure, and the small scale of the urban mesh that was required to serve the rural countryside even in the most elementary way.

A similar cellular structure, its units combining a walled city with a tributary area of farmland, has characterized China for more than three thousand years, and only within the last century has it been appreciably modified in some areas. The interior colonization of China from the homeland of Chinese civilization in the north-west was accomplished by multiplication of such cells. As the agricultural area was extended so new cities were added, the walled cities being as integral if not as numerous features of the humanized landscape as the agricultural villages. The city lived upon its surrounding farmland, but an important contribution it made in return to at least part of this area was the much-prized nightsoil. This traffic is still a noteworthy feature of town-country relations in China.

The rise of nation states in Europe, by extending the units

of political organization, was accompanied by the growth of their capital cities as they were enabled to draw upon the surplus of an extending area. Especially was this so when it was fostered by a deliberate policy of centralization. This was so notably the case at Paris that during the reign of Louis XIV it attained half a million inhabitants. Simultaneously, the development of ocean navigation and exploitation by Europe of the wealth of overseas lands gave the leaders among ocean ports new horizons of affluence and expansion. The new orientation of trade upset the old hierarchy of towns and established new urban values. Atlantic ports, such as Antwerp and Amsterdam, were in the vanguard with London, but those of the Mediterranean, which was becoming a backwater, languished. Political and economic factors combined to swell the population and enhance the importance of London. During the reign of Elizabeth its population doubled, and during the course of the seventeenth century it grew further from 200,000 to nearly 700,000, surpassing even Paris.

Only cities specially favoured by geographical position and historical circumstance to become the instruments of state-building or maritime expansion attained such stature, and they far outstripped their nearest rivals in their respective countries. In spite of the rise of Bristol, the relative importance of London among English towns, as of Paris among French towns, was greater than today. Late in the seventeenth century the two largest English provincial towns, Bristol and Norwich, each had only about 30,000 inhabitants. York and Exeter were probably the only others with as many as 10,000. London was indeed considerably more populous than all other English towns put together.

CHAPTER II

THE ORIGIN AND BASES OF TOWNS:

II. MODERN TOWNS

THE first urban revolution, as we have seen, was the product of agricultural increment and its amassment in favoured localities. It accompanied accessions to human culture that required for their full development specialists and their concentration in urban groups. After the appearance of an agricultural surplus adequate for the existence of towns, however, the ratio of food-producers to non-producers remained relatively stable for a long time. Man's limited command of power made impossible any drastic reduction in the proportion of food-producers needed to support society. Under such conditions the general importance of urban phenomena and still more the size of the largest towns depended upon the degree to which military conquerors could lay hold upon the economic surplus and organize its concentration in the seats of power, such as imperial Rome, or upon the activities of merchant capitalists in exploiting the opportunities for canalizing trade through particular channels and levying tolls for their services as middlemen as in the case of Venice. Both were essentially achievements of organization, and urbanism flourished or declined according to whether the small surplus was diffused or concentrated. Society showed different patterns of crystallization into towns as the political authority of individual states waxed and waned and the channels along which trade flowed changed. But human society as a whole was not urban; city life was not the norm.

In modern times, however, a second urban revolution has taken place. The vast accession of power which marks modern man's command over the physical world has enabled general urbanization. The old equilibrium between food-producers and the rest of society has been broken down, and this in turn has been accompanied by technological developments that have tended to direct a greatly increased surplus to the support of larger and larger urban groups.

It is axiomatic that since townsfolk must be supported by the surplus food production of agricultural areas the upper limit to the urban population of the world is fixed by the food-producing capacity of the rural population and by the possibility of transporting their surplus to the towns. Before modern times agricultural technique, especially in its dependence upon human and animal power, severely restricted the farmer's capacity to produce food in excess of the immediate needs of his family. Only in localities blessed by nature could the surplus be considerable, and elsewhere virtually subsistence economies prevailed, as they still do in countries relatively unaffected by western culture.

The one country where reliable statistics throw light upon the population structure in an agricultural society before it was much modified by modern transport and industry is India. The early Indian censuses from 1872 to 1901 show that up to the beginning of this century the rural population, directly supported by agriculture, amounted to quite 90 per cent of the total population. We may safely assume that a very substantial majority of human beings whose lives were organized to support the cultured life of the great cities of the ancient and mediaeval worlds must have been employed directly in agriculture. The first duty of the land on which they lived was to maintain them at least at the level at which they could work. Beyond this the surplus per family engaged in agriculture could not have been great. The depressed or enslaved masses of the rural countryside were the grim and sombre human background of the brilliant city cultures that flowered in ancient times.

It must further be remembered that so long as land transport was undeveloped there were great obstacles to the physical concentration of such scattered small surpluses. Where advantage could be taken of water-transport these difficulties were to some extent obviated, and the road system which the Romans flung across their empire enabled them to achieve wonders of political and economic organization. Nevertheless the extreme technical backwardness of ancient civilizations is often overlooked. It certainly limited urbanization. As Professor A. H. M. Jones has emphasized with regard to Rome, the methods employed in agriculture, transport, and manufacture were extremely primitive.

"Despite a crying economic demand the Roman world remained barren of any technical improvements to production and transport."

It depended essentially upon manual labour and the burden of supporting large classes of rentiers, bureaucrats, and soldiers, not to mention the urban proletariat, was more than this primitive economy could maintain. Jones concludes that a major factor in its decline was shortage of manpower and the failure to find an effective substitute. The same fundamental limitations persisted throughout the Middle Ages, and not until modern times did human inventions unleash a new potential of general urbanization.

Now, with the growth of science and its application to give man a new and increasing degree of mastery over his natural environment, the previous conditions of stability in the relations between urban and rural populations have been swept away. Beginning with the great advances in agricultural technique in the eighteenth century which were contemporary with the early phases of the Industrial Revolution in Britain, the old-established countries of the western world, with little opportunity to extend their cultivated areas, have vastly increased their agricultural production by improved methods. These include dispensing with fallow by use of scientific crop-rotations and fertilization, the introduction of new crops and of new and better varieties of established crops, the improvement of livestock by selective breeding, and the application of power-driven machinery to farmwork. Most of the changes have increased the output of food per unit area of farmland without any commensurate increase in the manpower employed. Indeed the net result has been a progressive diminution of the agricultural labour force. While the community has been rapidly increasing in numbers, the proportion of its members required for food production has been radically reduced.

Agricultural changes in the homelands of the peoples of the western world who have been experiencing urbanization, however, is only half the picture. At the same time there has been a great extension of the cultivated area of the world, brought about by the addition of new lands, far distant from these markets. The development of transport, both by sea and land, has given the populations of old countries the opportunity vastly to extend their field of supply, and other technical advances, such as the development of refrigeration, have further contributed to this trend. The coming of the railway has been the most spectacular of the developments that have opened up for food producing great areas in the interiors of continents that were previously out of range. Here by lavish use of easily occupied land and exploitation of the stored-up fertility of virgin soils, a small personnel, increasingly equipped with machinery, has been able to produce great quantities of food and to supply this food to distant markets by rail and steamship. The enormous extension of the area available for food supply made possible a prodigious growth of population and its concentration in towns in countries such as Britain. This expansionism, the keynote of the nineteenth century, refuted the gloomy theories of Malthus, at least for a time.

That the great increase in population which has been the most fundamental and far-reaching fact of human history in modern times should have come to be largely urban in its distribution, so as to destroy the old relation between urban and rural, has been due not only to the operation of factors which have released the rural arm of the balance. Powerful concentrating factors have operated to induce the urbanization thus rendered possible. Freed and displaced from the countryside, men were being drawn into towns because of their rapidly increasing importance as workplaces. So long as industrial production depended primarily upon manual operations, or upon small-scale wind- and water-power, manufacturing was not necessarily urban. From its earlier mediaeval association with trade, and therefore with towns, even such manufacturing as was geographically specialized tended later to scatter to the countryside. There it found freedom from irksome gild restrictions, in some cases special site requirements for its processes, but above all the reservoir of labour provided by families settled upon the land. It might continue to be organized from towns, but the actual production was much more an activity of the countryside.

Once steam-power was applied to manufacturing processes, however, a massing tendency was immediately apparent. The factories in which the new machines were installed needed to be concentrated, often in an extreme degree, though not necessarily in the old towns. The labour engaged in manufacturing operations which was thus abstracted from its former agricultural associations was concentrated in units so large in size as to ensure the presence of considerable aggregations of people wherever manufacturing was localized. Some were important accretions to old-established nuclei, others were new towns, which only tardily acquired the appropriate organs of urban government. It was this divorce of manufacture from primary production which in the early phases of the Industrial Revolution drew population from the countryside into the towns. Decline of rural handicrafts was an effective cause of rural depopulation long before agricultural depression set in. And not only was geographical specialization worked out within the individual countries of the western world, it went so far as to produce the familiar contrast and reciprocal trade between countries predominantly concerned with manufacturing and with primary production.

A century ago, at the time of the Great Exhibition, Britain presented a society in process of rapid urbanization, but nearly half the population were still enumerated as living outside towns, and a much larger proportion were country born. In no other country was the proportion of town residents anything like as high, but the typical Englishman was not yet a townsman. The industrialization that was massing more and more of the nation into towns had already created a numerous class of distinctively industrial towns. Industrial towns had existed before the Industrial Revolution, but as with other of its salient features such as factories and capitalism, it was not the phenomenon which was new but the scale on which it was developed. Specialized industrial communities were not unknown even in the ancient world, but they were quite exceptional. Never before had industrial towns become so typical of a society, never had they multiplied so amazingly or attained such dimensions of individual size. And as time went on industrialization spread to the Continent with similar urbanizing effect, although there its full impact came after the provision of a railway network and the factories were attached much more to established, older towns.

The modern organization of industry, however, has not been the only factor contributing to increasing urbanization. Equally important has been the shift from employment in making things to that in performing services. The development of the machine has not only released a larger section of society from food-production, it has also led to a general reduction in the proportion needed for direct production of material goods. The more advanced a country becomes, the greater the proportion of its people engaged in performing services instead of in making goods. Production now employs only half the working population of Britain; the other half earn their livings buying and selling goods, transferring them from one place to another, or discharging various kinds of personal service, as in professions, entertainment, the transport of an increasingly mobile population, or administering the greatly expanded social services. Since many of these services are centralized, their personnel is concentrated especially in towns, which are increasingly developed as service centres. In a typical balanced English town, service occupations of one sort or other employ a majority of the gainfully occupied population.

The disparities still evident between countries whose economies have been transformed by the adoption of modern technology and those where a peasant tradition persists illustrate the measure of this powerful urbanizing influence. Whereas in the working populations of Great Britain and the U.S.A. non-producers are now in a majority, in Yugo-Slavia 90 per cent are still engaged in production (nearly 80 per cent in agriculture) and only 10 per cent in services. The figures for Poland are very similar, and if comparable figures were available for

China they would doubtless show an even greater preponderance of producers. The switch from productive to service occupations that has taken place in modern times in the western countries has also been accompanying the more recent transformations of the Japanese and Russian economies. It is much less complete there, but is in rapid progress. Whereas in 1913 only 10 per cent of the Russian population were engaged in services, by 1939 the figure had exceeded 18 per cent. Australia has nearly as high a proportion in services as Great Britain, and in the U.S.A. the process has gone even further than in this country. Where Great Britain is distinctive is in the much smaller importance of agricultural workers among its producers; in respect of service and productive sections of the total population Great Britain, U.S.A. and Australia are substantially alike. They differ from countries such as France and Germany, where the shift from production to services is considerably less and about two-thirds are still employed in production.

Such has been the transference of emphasis in employment from production to services that even countries specialized upon primary production, where that is highly organized commercial farming engaged in producing for overseas markets, have become highly urbanized. The numbers engaged in primary production (less than 20 per cent in Australia) are only a small proportion of the total workers, and the rest find employment chiefly in towns, and in large ones at that.

Not only is the service element thus a generally prominent feature of the occupational structure of towns, specialized types of towns, notably holiday resorts, have arisen to discharge particular service functions. Their origins date back into the pre-modern phase of urban development, when resorts with urban characteristics were created by aristocratic patronage as centres of fashion and social life. It is since the railway age, however, that their numbers and size have greatly increased, as they have multiplied and grown to cater for the recreation of an enlarging sector of society, provided with increased leisure and the wherewithal to enjoy it. A concomitant of the intense industrialization and urbanization of Great Britain, an island country with stretches of seaside now within easy reach of all major concentrations of population, has been the development of resorts on a scale unmatched in any other country. As a class of towns, resorts are here both more numerous and more highly specialized in function. The cult of the seaside dates from the middle of the eighteenth century, when emphasis began to shift from the curative effects of taking the waters at spas to the benefits of sea-air and sea-bathing, advocated by the medical profession and made fashionable by royal example. The peace and security that followed the Napoleonic wars made the coastline safe for settlement, but it was the coming of railways that gave the main impetus to the mass seasonal movement to the seaside which is now such a feature of British life.

The numerous seaside towns other than ports that have sprung up round the coast of Great Britain, mainly during the last hundred years, are themselves susceptible to classification. In different degrees they are residences of retired people, dormitories for people who travel to work elsewhere, and holiday resorts for visitors who belong to different social strata and income-groups. Some resorts are fashionable and exclusive, others plebeian and popular. These differences are not only reflected in the relative prominence of index classes in the composition of their resident population as revealed by the census, but are clearly apparent also in their external aspect. Their physiognomy in terms of types of buildings and urban fittings, no less than the life of their streets and seafront as it changes with the seasons, are expressively varied.

Again, the modern application of large-scale organization to some forms of primary production, notably mining and fishing, has brought about a geographical concentration of these activities, which were earlier scattered. Considerable communities exist in the form of specialized mining or fishing towns, and these are equally distinctive in character and appearance. Wherever conditions have led to the concentration of large numbers of people, that is, wherever a large-scale activity is geographically concentrated so as to give a localized basis for mass employment, a town or at least the semblance of a town is created.

Thus the bases of urbanism have been greatly extended.

and it is possible to recognize many distinctive types in respect of function. To the categories already mentioned, and among which even more highly specialized forms may be distinguished, there are the other major classes of transport towns, such as specialized ports, packet stations, railway towns; and of garrison towns, such as military camps and naval bases. Nor does this by any means exhaust the possibilities of functional classification.

In size, appearance and form of government, any of the more highly specialized communities just referred to may rank as towns, although in developed urban functions and particular institutions they may be remarkably deficient for their size. Lewis Mumford, in *The Culture of Cities*, draws an important distinction between the mere massing of population and buildings and the social formation of a town or city to which it bears a casual relationship. He has stigmatized the nineteenth-century town as a simple combination of work-place and housing, in the form of factory and slum, multiplied in number and extended in area indefinitely without acquiring more than a shadow of the institutions that characterize a city in the social sense—a place where the social heritage is concentrated.

On the other hand, an urban culture introduces many features of its way of life, as well as its forms of building and its characteristic institutions, into the countryside even where the scale of concentration is not always urban by usual standards. That the residential development of an urban people manifests itself in urban forms is well illustrated by the Roman watering-places, and by many agricultural colonies from classical Greece which retained the essential characteristics of the Greek city. The terrace, which during the eighteenth century became the typical form of construction of urban residence, lent itself so readily to cheap mass production of housing that it lost its Georgian character of dignity and architectural distinction and was debased by jerry-builders. During the nineteenth century the terrace-form was applied wholesale to the problem of housing industrial workers alongside factory and pithead, wherever these might be. The setting was immaterial.

Today in Britain the process of general urbanization is so far advanced and has led to such an extension of urban forms in the settlement pattern that it may well be asked in what sense the term 'town' is applicable. What, if any, is the distinction between town and country that has contemporary significance? In what way can the epithet 'urban' be restricted in its connotation to express a really distinctive quality? It must be appreciated that there are distinctive sociological and geographical standpoints from which the problem may be examined. A town may be regarded first and foremost as a community of people pursuing a distinctive way of life as compared with the rural population of the countryside, or it may be considered as part of the earth's surface differentiated from rural surroundings by a particular type of human transformation with buildings and other distinctive structures.

Thanks to modern communications the overwhelming majority of the British people, whether or not they reside in built-up areas, share an essentially urban culture, and social life is very largely focused upon town-centres. If we regard a rural community as one in which a majority of members obtain their livelihood directly from the land, in this occupational sense there are few villages left in England. On the other hand, if we require of a town that it should be a socially integrated unit, considerable built-up tracts fall short of townhood and are only pseudo-towns in which not by any means all the essential urban organs are represented. Certainly the suburbia that intervenes between areas of urban life and areas of rural life is a no-man's land, neither town nor country.

In our modern society we depend more and more upon urban centres as work-places and for centralized services, but modern methods of transport make it possible to reside farther and farther afield while still participating in these. This has been responsible for the creation of extensive suburban tracts of housing, and more recently has begun to inject a considerable element of townsfolk into rural settings. Because of the influx of this 'adventitious' population, the rural community in the census sense, that is the population resident in Rural Districts, which are simply areas without an urban form of local government, is no longer decreasing in Great Britain. In recent decades the rural depopulation which had been proceeding steadily for some time before the first World War has been checked. Between the 1931 Census and the outbreak of war in 1939 rural and urban sections of the population grew at the same rate, but subsequent growth has actually been more rapid in the Rural Districts, so that these, as at present constituted, include 19.3 per cent of the population in 1951 as compared with 17.6 per cent in 1931. 'Back to the countryside', however, does not mean 'back to the land', and the proportion of the rural population engaged in agricultural production continues to fall. The agricultural character of the residents of the countryside is thus becoming less and less pronounced.

For his particular purpose the geographer must regard as urban a particular man-made type of landscape. Yet even from this point of view, the distinction between town and country has been blurred. In mediaeval times, close as were the relations between town and surrounding countryside, the physical distinction between the two was sharp. It lay along the line of the town walls, beyond which there was little or no suburban development. Nowadays there is no longer either socially or physically a simple clear-cut dichotomy of town and country; rather it is an urban-rural continuum that presents itself. There is no definite point where rural ends and urban begins.

"The concepts are clear only as they apply to the two extremes of the continuum, that is, to the most urban and the most rural. The distribution is not really a two-fold one, in which one part of the population is wholly rural and the other wholly urban, but a graduated distribution along a continuum from the least urban to the most urban, or from the least rural to the most rural. Consequently, the line that is drawn between urban and rural for statistical or census purposes is necessarily arbitrary." (United Nations Population Studies No. 8, 1950, p. 2.)

So inevitably are the distinctions introduced on population maps.

B

In different countries the arbitrary line between urban and rural is drawn differently, invalidating the comparability of the census data. The census statistics of most countries are not concerned directly with distinguishing urban clusters or agglomerations in the texture of settlement, but are content to separate the populations of unit areas which for purposes of administration are regarded as 'urban' and 'rural' respectively. While in France a *commune* is regarded as urban only if it contains a clustered population of at least 2,000 at the centre, in Great Britain and many other countries the basis of distinction is entirely one of the form of local government. In England the urban population is that which is resident in Boroughs and Urban Districts. There is also wide difference in practice from country to country in the recognition of urban status in this legal and administrative sense, so that whereas in Canada some places with as few as 200 people rank as urban, in Japan the smallest include populations of more than 20,000. The boundaries of the incorporated place, however, may be effectively limited to the built-up tract, and may even exclude some suburban extensions that might well be classed as town rather than country; or again, boundaries may divide a continuously built-up area into several municipalities. On the other hand, the municipalities of Japan (*machi* or *cho*) often include more than one urban nucleus as well as considerable tracts of agricultural land with villages.

To complicate the matter still further, in countries where the census distinction is based upon a classification of population clusters there are wide variations in the minimum size and character of those which are accepted as 'urban'. In Cuba all population clusters are so classed, though many are in fact hamlets or villages of fewer than fifty people, but in India the minimum size of a concentration that is regarded as urban is 5,000, although exceptions are made to include some smaller places which have 'other characteristics regarded as definitely urban'. In yet other countries the urban population is treated as that of places which serve as seats of administration, although in some of the countries where this applies it means that very small centres, with fewer than 100 inhabitants, appear as 'urban'.

These details have been given at some length because it cannot be too strongly emphasized that direct comparison of different degrees of urbanization by reference to the census figures of different countries is impossible. The recommendations of the United Nations Population Commission in 1949 refrained from attempting to establish a definitive distinction between urban and rural, but suggested the adoption by national census authorities of a classification of population clusters into a series of ten size groups, to be expressed for summary purposes in three classes, viz.: (a) dispersed, or in clusters smaller than 2,000, (b) in clusters of from 2,000 to 10,000, and (c) in clusters of more than 10,000. It was expressly advised that the terms 'urban' and 'rural' should not be used to designate the categories, which represent arbitrary breaks in the urban-rural continuum.

Within this continuum, however, there are nuclear areas where central services are concentrated, and which are the nodes of circulation. In a special sense, therefore, these are towns, distinguished from the intervening areas where the differences are simply those in density of buildings and degree of completeness of the transformation into a built-up area. Although the more open countryside that surrounds the densely built-up tracts consists of farmland that is primarily devoted to specialized forms of production in response to the nearby consuming markets, modern concentrations of population do not depend for their existence or size upon locally produced food. Released from the old-time relation to a surrounding territory which sustained it with food, the modern town finds its territorial basis as a service centre. The provision of many of the services which have multiplied as standards of living have risen and social organization has become more complex shows a relationship to area as well as to population. The location of a central service depends upon its accessibility to the population who are to use it; they must be within range. But within the field of the service centre there may be considerable variations in the balance between 'urban' residents living in built-up areas and extra-urban residents of villages and countryside. It is at points of concentration of central services within the texture of settlement that we may recognize developed urban functions and the essential character of modern towns.

For British towns we have investigated the group of key-services which taken together may be regarded as indicative of full urban status. Equipment as a shopping centre with a full range of specialized retail services, including branches of multiple businesses, is reflected by the presence of a group of three or four banks and a Woolworth's store. The latter may be regarded as an exceedingly valuable single gauge of shopping importance. If to these indices of economic status we add others representative of equipment as a district centre for entertainment and for education and health services—cinemas, secondary grammar school(s), and hospital—we have the hallmarks of townhood. The services specified are representative of the wide range of central services and institutions which are associated with developed urban functions and which hang together as a trait-complex.

A major feature of the current regional geography of urbanized countries such as Great Britain is the existence of conurbations, great tracts of built-up country. Whether by a simple process of outward growth, as the sectors between spreading tentacles become filled in, or by coalescence of a number of originally close-set nuclei as these acquire accretions, a brick and mortar unity is imparted to large continuous areas. The more or less continuous sprawl of urban structures, with only relatively minor enclaves of land unappropriated for urban uses, may or may not be recognized as an urban entity in the pattern of local government. Extension of civic boundaries usually lags behind the advance of the built-up area. In the case of fusion of once independent nuclei, municipal pride and jealousy have often retarded administrative recognition of the physical unity that has been produced. However obscured statistically by administrative partition, the geographical reality of these conurbations is incontestable. Their physical continuity can be appreciated by the traveller as he moves through them, and their unity is at once apparent from their appearance on large-scale maps or still more when they are viewed from the air. As Professor Fawcett has written:

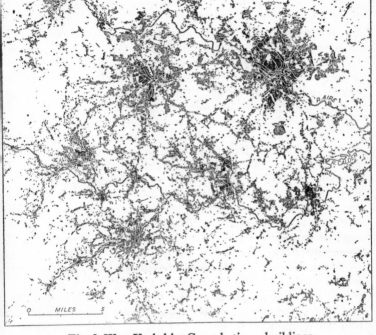

Fig. 2. West Yorkshire Conurbation—buildings

Nearly 1,750,000 people live in the area shown, divided among six County Boroughs, six Municipal Boroughs and more than twenty Urban Districts

"An observer in an aeroplane hovering above one of these conurbations on a clear, dark evening, when all its lamps are lit, would see beneath him a large area covered by a network of lights, glowing here and there in brighter patches where its main roads meet in its nodal shopping districts and elsewhere shading into the darker patches of its less fully built-up areas—parks, water-surfaces, or enclaves of rural land. To such an observer the continuity of the conurbation would be the most salient fact about it."

Whereas at the dawn of the nineteenth century London

was the first urban concentration ever to exceed a population of a million, and remained alone in this millionaire class until Paris joined it about 1850 and New York about 1860, there are now some eighty such giant conurbations in the world. Of these seven are in Great Britain, where Greater London now includes between eight and nine millions according to variant delimitations of its extent. Many more conurbations exceed half a million inhabitants, their number and sizes being understated by the official figures which recognize only administratively defined towns and cities.

The concentration of urban populations into giant aggregates is a striking feature of contemporary urbanization. In part it reflects the modern tendency for the geographical structure of modern society to become not merely urban but metropolitan. In Great Britain the seven largest conurbations account for almost 40 per cent of the total population and for half the town-dwellers; in Australia four of the six State capitals concentrate more than half the total population, and three out of every four town-dwellers live in the capital cities.

Gone for ever are the stable relations of the Middle Ages between urban and rural sections of the population, and a new position of equilibrium has not yet been attained. It is true that in Great Britain where the second urban revolution first began and where the process has gone farthest, there are signs of a new stability arising in the relations between urban and rural, classified by residence. But the urban population of the world generally is still growing much more rapidly than the total population, and recently the full impetus of the trend towards urbanization has extended to the U.S.S.R. and to the East. Even in China, where probably well over three-quarters of the population are still rural, living essentially in agricultural village clusters, there are now seven cities of the million class. The largest, Shanghai, with 4,500,000, is entirely a product of the last hundred years, as are several other great commercial cities.

In Japan the process of urbanization has gone much farther. Most of the amazing increase of population since 1860, which has carried the total from the previously stable figure of 30,000,000 to treble that number, has gathered in

MODERN TOWNS 39

towns. The increase in recent decades has been very largely
drawn to the four great urban and industrial nodes that are
strung out from east to west from the Kwanto lowland to the
new industrial district of northern Kyushiu. Tokyo-Yokohama,
with more than 8,000,000, and Osaka-Kobe, with more than
5,000,000, are great bi-nuclear conurbations that rank with
the largest of the West.

In the U.S.S.R. the transformation of Russian society by
the Communist revolution has also been accompanied by an
urbanization which shows gathering momentum. Between
1926 and 1939 the urban population, defined as that living in
clusters of more than 500 with an urban form of government,
more than doubled (from 26,000,000 to 56,000,000) while the
total population only increased from 151,000,000 to 171,000,000.
The number of conurbations with more than 100,000
inhabitants had increased from thirty-one to eighty-two while
the largest, Moscow and Leningrad, now have more than
4,000,000 and more than 3,000,000 respectively. In a study of
population of the Soviet Union published by the League of
Nations in 1946, Lorimer drew attention to the appearance of
a remarkable number of new towns of considerable size. There
were in 1939 no fewer than forty-nine 'boom cities' of 50,000
or more people, which had more than trebled in size during
the preceding twelve years. Some of them, such as Karaganda
and Magnitogorsk, towns of the order of 150,000 population
in 1939, had not existed at all in 1926.

Against the time-span of human history or even of Western
civilization, the period during which population has increased
rapidly is short. Not until 1750 did the gap between birth-rate
and death-rate begin to widen significantly so as to bring
about the amazing increase of population in the Western world
which accompanied its acquisition of modern science and
technology. A falling birth-rate has closed the gap again, and in
the West the episode of rapid population increase has come to
an end, limited in time to a mere two hundred years. But it is
still in an early stage in other human societies where the
impact of Western technique is only now being felt to the full.
All the evidence of present trends goes to suggest that as the
upward surge of population, to which the West is no longer

making significant contribution, is worked out in these new fields it will likewise be accompanied there by large-scale urbanization. The extra world population in the next century as in that which has passed will be gathered largely into towns; but they will not be people of stock derived from peninsular Europe.

CHAPTER III

THE SETTING OF TOWNS

PHENOMENA which other sciences isolate for systematic study are examined by geographers in their associations in area. This geographical point of view is concerned with more than the facts of distribution, which of course are characteristics that properly form part of any thorough systematic studies. Geography concentrates attention upon their settings. In thus describing the geographical context of towns it is desirable to distinguish different aspects of position, which may be designated location, site, and situation. To the question "Where is it?" each offers a different and in itself only partial answer.

Location can be stated quite tersely and precisely in terms of latitude and longitude, or distance and direction from other established points. The co-ordinates of a recognized grid reference system define it precisely. It is a specific, unique attribute of any town or place, and is the primary information given by a gazetteer. In geographical description locations correspond to dates in historical narrative. But geography has moved beyond gazetteer description, and is no more a mere catalogue of locations than the work of the historian is a record of events as lists of dates. Just as the historian exercises his critical judgment in writing the chronicle of events by drawing attention to the significant relationships between the happenings he associates in a time sequence, so the geographer is concerned with selection of the significant groupings of phenomena in area. In setting out thus to portray the *milieu* of towns the urban geographer must describe the site, which may be defined as the ground upon which a town stands, the area of earth it actually occupies.

Although no two towns have sites that are exactly similar, it is not difficult to recognize well-defined categories of town-sites. Certain physical features, for one reason or another, have been favoured for the siting of towns, and provide a basis for classification of towns according to site types. The site is enlarged in the process of urban growth, yet it nevertheless remains an area local and relatively restricted, and as such is only part of a much wider setting which affects the origin and growth of urban characteristics. This brings us to the conception of the situation of a town, its position in relation to its surroundings. Some elements of this wider setting are altered in the course of time, and others, while remaining permanent features of the scene, change in their significance for the life and development of the town. As with the sites of towns, so with their situations, the geographer is led to recognize categories and towns are commonly and usefully classified on this basis.

To describe a town as a gap-town, however, ought to imply more than a mere statement of one single fact among the many that might be made about its physical setting. It is a merited appellation, summarily describing the situation, if it draws attention to something that has been especially significant for the urban development and functions. There can be little value in the use of generic terms unless such depth of meaning is implicit. Even so the term may be a very loose description except when specifically related to a particular phase of urban development. In the North and South Downs respectively, Guildford and Lewes are gap-towns in that they control important passages used by modern routes through these ridge barriers. But long before these transverse routes became important both towns experienced a long phase when the significance of their situations rested in their control of the crossings of rivers and their alluvial valley-floors. These lay athwart the longitudinal routes running west-east along the belts of dry, relatively open country afforded by the chalk ridges amid the forest-choked lowlands. The designation 'gap-town' is thus misleading in reference to their origins, since it focuses attention upon a later significant feature of their space-relations and obscures their geographical *raison d'etre*. The physical features that presented the obstacles, for the negotiation of which they occupied significant situations at first, were rivers and valley bottoms, not upland ridges. At

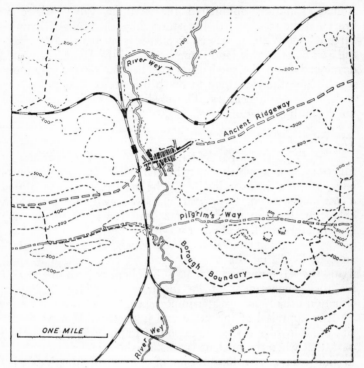

Fig. 3. Guildford

Within the modern borough boundary only the plan of the pre-eighteenth century town (unwalled) is shown, and not the extensive built-up area of today

Guildford the 'ford' element in its name, and the west-east alignment of the town along its axial High Street, bear witness to the feature that was long dominant. Not until the coming of the railway in 1845 did the north-south line of movement through the Wey gap become of primary importance.

At Lincoln, although there was important waterborne traffic through the Witham gap from early times, the full significance of the gap for transverse movement was only realized after the construction of the railways. The nucleus,

Roman Lindum, dominated the Witham valley where it breached the south-north limestone ridge that carried Ermine Street. The Witham waterway offered easy penetration from the east coast to this land route here, and the Danish borough developed from the nucleus of the Roman castra, extending south along the ridgeroad to reach the site of Bargate by the end of the tenth century. The mediaeval city had a definite north-south elongation.

The Sites of Towns

Towns grow in particular places to discharge necessary functions, among which it may be that one is of primary importance, so that it may justifiably be regarded as the *raison d'être* of the town. It may of course happen that the urban character, both in respect of size and function, emerges by growth and accretion about a pre-urban nucleus. In each case, however, it is the conditions of site which have special importance in localizing the original function at a particular spot, fixing there the nucleus. For its subsequent growth in size and for the enhancement of its function the wider setting or situation usually has greater importance.

Any appraisal of the value and importance of a particular site must involve a reconstruction of the geography of a time past, that when the nucleus was established. Unless considered in this context the localization of the urban functions may often appear capricious and meaningless, not to say absurd. There are many sites occupied by established towns which, in spite of all the resources of modern engineering, could not be conceived of as possible choices for new towns if they happened to be available now as virgin sites.

Among site considerations that have played a part in fixing the scenes of urban development ease of defence has often been paramount. In innumerable cases a nascent urban community has gathered under the shadow of a castle, for which in turn a naturally defensible site had been chosen. Such is the significance of 'burgh', which in its variant forms is so common an element in town-names. Even where the strong point itself is missing, the olden town usually sought a site naturally endowed with obstacles to free entry by attackers, and the

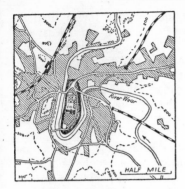

Fig. 4. Durham

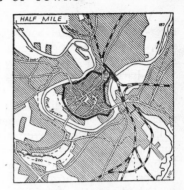

Fig. 5. Shrewsbury

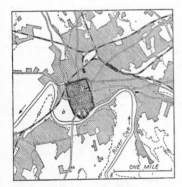

Fig. 6. Chester

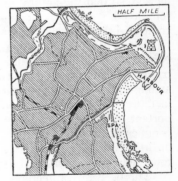

Fig. 7. Scarborough

The thick line in Figs. 4, 5 and 6 shows the position of the mediaeval walls enclosing the urban kernel (cross-hatching) within the modern extensions of the built-up area (single hatching). Railways are shown distinctively.

function of town-walls was to strengthen and complete natural defences. Steep slopes and water barriers, singly or in combination, especially offered such defensive facilities. The acropolis, the city set upon a hill, epitomizes a familiar geographical association, well-nigh ubiquitous in the Mediterranean world, and very characteristic over much wider areas and during long periods of urban history.

Alternatively, in the case of innumerable towns sited on river-banks, the water-barrier gave protection on at least one side, witness the significance of the general siting of the riverside towns founded by colonizing peoples with the water-barrier in front. This is exemplified by the urban foundations of the mediaeval German colonization east of the Elbe, with their predilection for the west or left banks of the rivers. In the European penetration into America from the Atlantic seaboard sites on the east sides of streams were correspondingly favoured. Sites in meander loops, such as Shrewsbury (Fig. 5), in confluence forks, such as Lyons, Coblenz and Pittsburgh, or in other types of peninsula enjoyed special advantages of water protection; and even more favourable sites for defence were provided where the streams were incised, so that the natural protection of the water-barrier was reinforced by a steeply sloping approach. Promontory sites that combined these advantages were possessed in notable degree by Durham (Fig. 4), Besançon and Toledo. Sometimes, as at Newcastle-upon-Tyne, additional flank protection was given by tributary valleys (Fig. 17). The examples cited are only a few among many which in greater or less degree enjoyed such facilities. Another defence consideration that was always important was the availability of a water-supply within the confines of the site. Impregnability to direct assault would be of little avail if an inadequate local water-supply rendered the town helpless to withstand siege.

The need for protection, which induced towns established in times and areas of insecurity to seek such defensive sites, was not always easy to reconcile with facilities for discharging normal peaceful functions such as trade. For those towns which made use of or depended upon navigable waterways the latter were at once channels of approach for trade, and protections against land attack. The need for facilities to load and unload ships gave added importance to sites that could offer shelter from stormy weather and suitable water-frontage in contact with a deep channel. The mouths of tributary creeks in the main waterway gave such advantages, as the Walbrook at London and the Hull at its mouth in the Humber. The latter also exemplifies in marked degree, as does Bordeaux, the

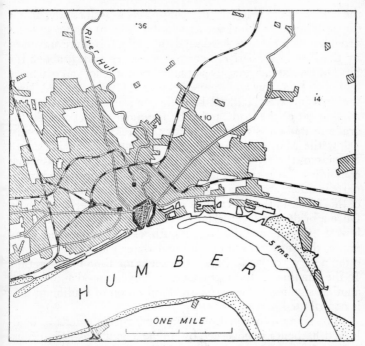

Fig. 8. Kingston-upon-Hull: the modern built-up area (single hatching) with its nucleus, the walled mediaeval town at the entry of the Hull into the Humber

importance of a site approached by deep water through the swing of the current to the concave bank at a major bend. The commercial advantages of a fairly deep, well-sheltered tidal creek, the Hull, easily approached by the deep, stable Humber channel outweighed such disadvantages as the prevalence of low-lying alluvium on which the town had to be built.

A large class of towns have had their original sites determined by advantageous conditions for crossing rivers, or their marshy valleys, which were sometimes more serious obstacles to land-routes than the streams themselves. In such cases the approach of firm ground to the river-bank offered both a well-defined constant channel to cross and opportunities for the land-route to reach the crossing-place, as well as advantages for building near the river. At Chester (Fig. 6) such a site was provided by a ridge of hard sandstone which is there traversed by the Dee in a relatively narrow and steep valley, above the head of the estuary and its salt marshes but below an extensive belt of wet clay lowland. Sometimes, as at Gloucester and Paris, the crossing was facilitated by the presence of islands. The adjustment of the load carried by the Seine to a change in the grade of its bed is responsible for the presence of islands just below the Marne confluence, and here at Paris an important crossing of the river was fixed. The prevalence of town-names containing elements such as '-ford' and '-bridge' or their counterparts in other languages, bears witness to the common operation of this factor in the selection of sites favoured for urban growth. Once a bridge has been established, more especially if it is the lowest bridge above the river mouth, as at London and Newcastle, the convergence of land and water routes is fixed, and often a very precise limit is set at the same time to navigation. In the handling of traffic, break of bulk is imposed and thereby another significant factor for the localization of trading groups is contributed.

The siting of original port functions within extensive land-locked harbours is governed by similar conditions—the occurrence of a stretch of shore combining facilities of access by land and water such as may be offered by a peninsula of firm ground jutting out towards the deep-water channel and offering at the same time a maximum of water-frontage. Such has been the case with the nucleus of the port and town at Boston. In Britain, Southampton, Plymouth, Pembroke and Harwich offer modified examples of fundamentally similar conditions.

Analysis of particular cases reveals innumerable variants and individual peculiarities in respect of the conditions that have determined the siting of towns which have developed as administrative or commercial centres, and the manifold combinations of operative siting factors. To single out from among them the most appropriate for the purpose of site classification is often difficult, and in any system equally strong arguments might often be adduced for claiming a particular town as a representative of different site-categories.

Towns distinguished by other specialized functions have likewise been affected in their siting by the conditions required by or especially appropriate for the discharge of these. Thus among industrial towns the factories of which they are extensions are often concerned with operations which use water in such large quantities that they have sought sites alongside streams. Some industrial towns have developed round factories which were originally set up to use water-power. They were consequently sited at breaks of slope along stream courses, which sometimes had further significance by interrupting navigation and compelling traffic to break bulk. In other cases, site requirements peculiar to the industries concerned have governed the placing of the factories, which in turn have been all-important for the genesis and growth of the town. Tracts of flat land adequate for the lay-out of the factories and for the indispensable transport services, though not always for the housing and other urban buildings, have also exercised importance. In districts where ruggedness limits the extent of such sites, the location of industrial development may be closely controlled in this way, as, for instance, it has been in Japan.

In the exploitation of some mineral fields, housing and urban services have been assembled at the sites occupied in an earlier phase of economy and settlement, as at Pittsburgh; but in others, where the mining settlement represented the original colonization, the occurrences of highly localized ore-bodies have themselves controlled the siting, as at Johannesburg on the Rand and Broken Hill in Australia. The significance of springs of water possessing special properties needs no emphasis in connection with the siting of spa towns. Among other resorts coastal features, such as attractive bathing beaches and easy access to the foreshore, have played their part, although site advantages have not in every case been very powerful. Considerable stretches of coast show little differentiation of site conditions, so that the advantages of one spot over another are not significant as compared with the over-riding influence of other factors that affect the incidence of urban development. As with knolls in tracts of flat or ill-drained land, springs in tracts of waterless country, and flat terrains in rugged country, it is where site advantages are localized, exceptional features, and not of widespread occurrence, that they become a decisive influence. In accounting for the siting of many seaside towns, the existence of an earlier nucleus of settlement to which the modern town is an attachment must be recognized. As the trade or fishing, which had been the mainstay of these not necessarily urban settlements, decayed during the course of the nineteenth century, they embarked upon a new career as resorts by developing adjacent stretches of the sea-front. Their nineteenth-century extension was dominated by the positions of the marine promenade and the railway station. Among many examples are Scarborough (Fig. 7) and Aberystwyth (Fig. 16) and, on a much greater scale, Nice. Fortified headland, small harbour, and adjacent foreshore are common site features of those named.

Only very exceptional towns are as much products of the resources of their own localities as Johannesburg. Even for spa towns accessibility for their clientele is quite as important as the springs, and this applies still more to the emergence and growth of holiday resorts. The more plebeian their character, the closer the physical association with the populous areas they serve. Special site attractions have more scope to influence the development of the more select resorts. In accounting for the origin, much more the subsequent growth of the vast majority of towns, the wider regional setting must thus be invoked. On the other hand, the influence of site conditions, seen to be so important for the localization of the urban nucleus and for the disposition of the constituent elements of the original settlement, is not limited in relevance to the elucidation of urban origins. Throughout all later growth the physical nature of the terrain affects the form of the urban development. This applies especially where the site is varied and where such major physical controls as steep slopes, water-bodies or marsh areas are strongly represented, but it obtains in some measure in every town. Both in the changing outline of its perimeter and in the re-organization of its internal pattern of land-use, the growing town reflects adaptations to the opportunities and limitations of its site.

Under conditions of deeply engraved relief, there is a predisposition to linear development, as in the long, straggling towns of the South Wales coalfield or the industrial settlements, including Elberfeld and Barmen in Westphalia, which have coalesced into the conurbation appropriately known as Wuppertal. The control of relief upon the development of the typically elongated valley town is characteristically exercised through its canalization of communications, and the linear extension of building along the dominant routes. The Wycombes are a notable example on a trunk route through the Chilterns. Equally expressive of a physical pattern, but one of a very different type, are towns of flat deltas, such as Venice and Osaka, where the network of distributary channels is fundamental to an interpretation of the urban morphology.

In the process of urban development, however, sites may be considerably modified. Stream courses may be diverted or carried underground, marshes drained and mud-flats reclaimed, and minor irregularities of relief such as small, steep-sided valleys may be effaced by dumping refuse till they are filled in. The present-day contours and surface drainage may thus fail to reveal features of profound significance for earlier phases of the town's history. The left-bank tributaries of the Thames, which were so important in the early development of London, like others farther west, exist now only as names—Shoreditch, Holborn, Fleet, Tyburn, Westbourne. The deep denes of the Lort and Pandon burns, major features of the early geography of Newcastle, have long ago been filled in (Figs. 17 and 18), and farther east the much larger valley of the Ouseburn is now suffering a like transformation below Jesmond. The marsh belts which so long confined London to the north side of the Thames, and maintained open fields around the gravel island of Westminster, like the *Marais* which confined mediaeval Paris on its north side, along an ancient meander loop of the Seine, have disappeared under modern buildings. The greater part of the centre of Belfast (Fig. 14) has been built upon reclaimed estuarine flats (slobland) at the head of the lough, its greater buildings such as the City Hall being supported upon piles. In New Orleans the surface hydrography of the level delta site, which in the past profoundly affected the lay-out of the growing city, is a factor which is no longer apparent, since for health reasons surface water has been almost entirely banished from the urban area.

Apart altogether from such modifications at the hands of man, there have sometimes been secular changes in site conditions, especially along coasts and in estuaries, that have profoundly affected urban fortunes. The eclipse of the old Cinque Ports and of Bruges by the silting of their channels of approach, and the decline of Chester as the Dee estuary became choked, are familiar examples among many that could be cited.

Site conditions that have been excellent for the original function or for a particular phase of urban development may become constraining, hampering features for the town later. In particular defensive sites, with their limited accommodation and restricted conditions of approach, are often ill-suited for urban development beyond a stage that is soon reached. Thereafter their further growth has been fraught with difficult problems. The more pronounced the natural conditions that once favoured defence the less likely is the site to be suitable for a modern town. Such towns labour under special difficulties of traffic circulation, unless the original nucleus has been left aside by the growth of a new town. This may be of old establishment or may date from the coming of the railway which was unable to negotiate the steep gradients necessary to reach an acropolis site.

Sites that have been quite adequate for the early development of towns may be outgrown, or under changed conditions may be outmoded. The increased size of shipping has caused once flourishing port towns to become moribund, and even major ports have needed to develop deep-water outports. Sites that are still ideal for the discharge of port functions may be highly inconvenient for the accommodation of urban development on the scale that is now engendered by the activities of a great port. It is so with New York, hemmed in to the west by the formidable physical barrier of the Palisades. The penetration of deep-water channels that so wonderfully endows New York as a great ocean port contributes formidable problems of congestion to the working of the vast urban organism. Nevertheless, it is another feature of the site, the composition of Manhattan Island and the western end of Long Island in Brooklyn by solid crystalline rock, that enables the serious problem of lack of space to be partially solved by the construction of giant skyscrapers.

In contrast, London has had far more space to spread outwards freely, and use of this space has been facilitated during the period of most rapid extension in the last half-century by the ease of constructing an underground railway system in the London Clay. New York is but one of many port-sites, endowed with long water-frontages for port installations, but severely cramped for building on tracts of land accessible to land-routes. Bombay, Boston, San Francisco and Portsmouth (Fig. 15), all on island and peninsula sites, illustrate the frequent incompatibility of enjoying optimum conditions for extension of port facilities and for the associated urban development. Other ports may be so hemmed in by mountains that accommodation of city as well as port is rendered exceedingly difficult, as at Marseilles. At Genoa such are the limitations of the site in this respect that the acquisition of the industrial functions so characteristic of most other great ports has been denied.

The site of a town is invariably an influence fashioning its development that is clearly discernible throughout its history. Its significance for the geography of the present-day urban area goes deeper and is more pervasive than those surface accidents which survive as obvious features of the urban scene. It is in the initial stages that advantages of site are most stimulating to urban growth and disadvantages most inhibiting. As time goes on, a momentum is acquired that makes continued growth less dependent upon site conditions, although the form it takes is always affected by them. No matter how favourable the site may have been for the initial establishment of the nucleus from which the town develops or for the earlier stages of this growth, there must sooner or later come a time when further extension meets site features that are handicaps. That towns do nevertheless continue to grow, accommodating themselves to these disabilities, is a measure of the importance of the capital already invested in essentially immobile equipment.

Nowhere more than in its urban manifestations does a culture show how important are its residues in reducing its freedom and resilience in the face of change. It is because towns especially are the fixtures of civilization, which cannot readily be improvised but which are built up by patient effort, that they are so persistent and capable of continued growth. Though an age may have some new towns, the community cannot on any large scale afford to sacrifice the old towns and start afresh elsewhere in conformity with current geographical values or social standards. Make do and mend is a fundamental principle in the history of cities as of civilization, and towns exhibit their adjustments to the difficulties as well as to the opportunities presented by their physical settings. Physical handicaps long felt or newly apparent are overcome, but only at a cost, which represents the operation of a law of diminishing returns that is sooner or later apparent in the development of every site where a settlement is established. The past weighs heavily upon the present.

The Situations of Towns

Settlements never outlive the formative influence of their sites. But whereas site is a controlling factor the importance of which towns share with villages, situation is peculiarly an urban attribute. Unlike rural settlements, towns are only exceptionally maintained by the intrinsic resources of their sites. They exploit the possibilities of concentration at a centre, and for this purpose depend upon the use of transport. The material bases upon which they live are external, so that situation governs their growth. And situation is a matter not only of disposition of physical features as these affect the canalization and concentration of traffic, but also of the political geography as it affects the extent of the territory to which the urban functions are related. We shall have occasion to return to this subject, but first it must be noticed how towns are related to routes. "Les routes ont fait les villes" was an aphorism of the French geographer, Vidal de la Blache, and certainly towns tend to grow at stage points along routes, and especially at the junction of routes or where break of bulk takes place. Towns are nodes of route-systems, and their importance closely reflects the degree to which they possess the property which has been called nodality.

Physical configuration, powerfully affecting route patterns, naturally plays a big part in endowing situations with potential nodality. Such natural nodality predisposes these situations to be scenes of traffic concentration and urban development. Thus the stage is set for urban activity where valleys converge in country of strong relief or at the junction of highland and lowland where channels of movement, concentrated along the valley lanes in the highlands, deploy more freely as the valleys open out into plain. Upon these gateways traffic from the plains concentrates to enter or penetrate the mountains; and exchanges, arising from the juxtaposition of contrasted environments with different products, enhance the commercial possibilities of such situations. Thus piedmont towns present a numerous and important class. Some of the major examples are found along the margins of the Alps in Europe and of the Appalachians and Rockies in North America. In Britain we may point to towns such as Shrewsbury along the Welsh border, and others such as Leeds, along the Pennine margin. Within highland areas, valley convergences afford nodality in minor degree, and urban functions are often developed in a graded sequence according to the measure of concentration of economic and social relations. Towns are strung out in descending size and importance from piedmont into the heart of the highland. Although the presence of concentrations produced by other localizations, such as those of mining or manufacturing, may complicate and modify this pattern, it is often clear.

Physical barriers, whether highland ridges, sea-inlets, rivers or marshes, induce concentration of routes upon the localities where they can be crossed most easily, and this in turn promotes urban development. In other cases, convergence of routes is brought about at the end of the natural barrier, where they bunch together to circumvent it. A classic example is the effect of the barrier imposed in the way of east-west land routes across the north of the U.S.A. by the southward extension of Lake Michigan, a fact which largely accounts for the nodality and importance of Chicago at its southern end.

Since water-transport has been far superior to land-transport in most parts of the world until modern times, signal advantages have accrued to situations where the penetration of tide-water inland allowed the maximum use to be made of sea-transport. Estuary-heads were more favoured than estuary-mouths, especially as the small shipping of earlier days was better able than modern craft to navigate the approaches. At the estuary-head a limit was set by the bridge that marked the lowest land-crossing, where land-routes converged to negotiate the water obstacle. Here, then, in a special degree, was presented the endowment of a situation with the priceless gift of nodality—where land-routes met in contact with the sea, at a break of bulk for water-borne traffic. Such in essence was the situation of London, and of numerous other port-towns. Like Glasgow, Newcastle, Rouen and many others, London belongs to this class of 'estuary-head' towns.

How critical for their rise was the fixing of the lowest bridge-point, a matter of siting, is abundantly attested by examples round the estuaries of Britain. Bridgwater, on the river Parrett in Somerset, leapt into prominence as a port and a town sprang up about 1200 when the local lord built a bridge as well as a castle there. The road from Glastonbury which had previously negotiated the river-crossing by Combwich ford seven miles downstream was at once diverted to this bridge, and at the same time access by sea-going vessels upriver to the older Langport was barred. Ultimately, transfer of inland traffic from the waterways to turnpike roads and later railways benefited Taunton at Bridgwater's expense, and the latter declined as a seaport as ships increased in size and overseas trade was concentrated upon major ports.

Although situation, expressed thus in terms of physical relations, is so fundamental for an understanding of urban growth and functions that the geographer is amply justified in endeavouring to categorize towns according to types of situation, a terminology at once significant in what it draws attention to, and free from woolliness, has yet to be developed. Gap-town, bridge-town, foothill or piedmont town are well *lyons*

established and acceptable examples. Less so is 'confluence town' unless applied strictly to the towns whose urban functions derive in large measure from a convergence of waterways. It is an ambiguous description of an urban situation where the nodality is due to the convergence of valley roads—and railways—even though the site may lie at the confluence of rivers. The crux of the matter is that the classification of site and situation should not be confused, as in attempting to telescope them.

For port development physical configuration, as it affects the orientation and concentration of inland routes upon particular stretches of coastline, makes situation of far greater consequence than local conditions of site. However magnificent a site may be for the approach and accommodation of shipping in a sheltered harbour, it cannot create a port unless the situation allows the development of the all-important relations with a hinterland. So powerfully may situation apply to promote urban development that towns may appear in areas in default of the existence of really favourable sites. This is perhaps most evident in the case of some ports for which favourable natural facilities are seriously lacking from considerable stretches of coastline.

A town may achieve great size and wealth thanks to the endowment of its situation, although its site may have little to commend it and may even be a persistent handicap. Great towns have arisen in many places in spite of serious drawbacks of site, because the situations demanded the presence of urban functions, and, as it were, called towns into being. Although situation may thus be a compelling influence that overrides the deficiencies of site, more usually it simply provides the stimulus for a degree of urban development somewhere within a more or less confined area in which the situation can be exploited. Local site advantages or even historical accident fix the precise spot. Potentialities of situation are seen to apply to an area, albeit more or less limited in extent, rather than to a point. It follows that the urban functions and development that derive from them may become attached to any of several sites where they can be realized. Which site emerges as urban may depend upon the established facts of an older phase of settlement, which may have been entirely rural and agricultural. Alternatively, it may depend upon the operation of such purely local considerations as defence. Nor must we overlook or underestimate the purely fortuitous element in deciding the spot. When it is remembered that so many factors can bear upon the matter it is not surprising for example that the precise topographic relationship of gap-towns to the gaps they command varies considerably. There is no standardized disposition.

The spot at which the possibilities inherent in a situation are realized in urban development may thus be fixed by local advantages of site or circumstance. These may determine the matter decisively, and more or less permanently, as at London. But it does not always happen thus. With changes in site value the inheritance of the advantages of a situation may pass from one urban site to another. The essential situation as a port town that Chester once enjoyed has been assumed in modern times by Liverpool. Roman and mediaeval Chester prospered at the head of the Dee estuary, in response to the opportunities for trade that were offered where the lowland gap between the Pennine and Welsh highlands opens a gateway north-west from the heart of England to the sea opposite Ireland. By the time progressive silting of the Dee had sealed the fate of Chester as a port, and its role had passed to Liverpool near the mouth of the neighbouring estuary, the full implications of the situation had changed. No longer did this north-west gateway face only Ireland. A new world of untold wealth was opening across the seas. With this outlook, and a hinterland which was soon to experience large-scale industrial development, it was a vastly enhanced situation that Liverpool exploited to her profit.

Offering certain situations the quality of natural nodality in special degree and denying it elsewhere, the physical lineaments of the earth's surface are highly significant for the growth and destiny of towns. In one situation urban growth seems natural, in another well-nigh inconceivable. Nevertheless, we must avoid a crude determinism that fails to grasp the importance of the intermediate links in the chain of cause and effect. Towns are the products not of gaps, valleys, bays and estuaries, but of man's use of these features for his routes. Nodality is imparted by the convergence of established routes, and is as lasting, but not more so, than the routes. If, for one reason or other, trade should leave its old channels, the life-blood of towns is withdrawn, and they decay. Urban fortunes are demonstrably dependent upon changing patterns of trade. The development of ocean shipping during the sixteenth and seventeenth centuries re-oriented trade, opening vistas of wealth and expansion to the Atlantic ports of Europe, and spelling decay for the Mediterranean ports which had flourished during the Middle Ages. Some of these, notably Marseilles, were later resuscitated by new developments such as the opening of the Suez Canal.

The towns along the Mississippi and its tributaries, which had been the chief urban centres in the phase of European penetration into the American interior, owed their prosperity to the trade that used the rivers. During the second half of the nineteenth century they suffered eclipse as ports, and also as towns, unless they were equipped to assume the new role of nodes of the railways that superseded the waterways. Similarly, the coming of railways in India launched some places in the Ganges plain upon a phase of vigorous growth as they acquired importance as collecting and distributing centres of the new system of communications. Cawnpore especially profited thus. But in by-passing old river-ports, the railways condemned them to stagnation and decay. A difference of a few years in the date of acquiring the railway was sometimes decisive for the establishment of these trends.

Only in the new countries where colonization has been a product of the railway age, achieved through the instrumentality of railways which were the pioneers in the wake of which agricultural settlement followed, is the urban pattern entirely or even largely a creation of the railway network. Elsewhere, railways came to forge new links between the principal established towns, though in so doing several minor centres might be left aside. Once established, towns attract routes, and in old countries like Britain and western Europe the railways tended to follow the same pattern as the roads, so that the nodality of old-established towns was confirmed and emphasized.

Even in the American Middle West a great many features of town distribution go back beyond the railway age, though railways have provided many new knotting points at junctions or places which for a time were railheads, so extending and filling in the older urban pattern. Yet even in the old countries the railways did not conform to the older pattern in every detail, partly because they were so much more sensitive to relief control, partly because of the prejudices of landowners and municipal rulers. Most towns, however, showed themselves keenly alive to the benefits railways could confer, and vied with each other in seeking railway connections. Equally it was true that the railway promoters sought the trade offered by prosperous urban centres so that the positions of established towns guided the courses of the railway lines. Nevertheless, some new nodes were created, budding points for new towns, such as Crewe; and some old towns, by-passed by the new arteries, were left to stagnate.

Comparison of the market-towns of eighteenth-century England with the service centres that are fully-equipped towns today shows that most of the old market-towns which have failed to maintain their status and acquire the modern organs of townhood are places that were left aside by the railway system. Before new life could once again course along the roads that converge upon them, a new and more centralized social and economic organization had been established, and revival was not for them. Even places such as Shepton Mallet which had to wait a decade or two for railway connections lost ground appreciably as compared with towns served by the earlier railways. Usually the delay was itself a measure of a decline that had already set in, and delay or failure to obtain the essential railway link in turn emphasized the decline. To an important extent, therefore, the changes in an urban hierarchy brought about by new media of transport amount to an acceleration of trends already started.

The immediate effects of new and improved communications, illustrated alike by the provision of turnpike roads in Britain before 1830 and by that of railways subsequently, is to promote the prosperity and growth of the established towns that they seek to serve, and to seal the fate of others that lie off their direct tracks and are too poor to claim connections with them. In course of time, however, other consequences appear; the concentration made possible involves the selective growth of the urban centres that are best fitted to profit. These extend their range and intensify their grip, while others become increasingly redundant. This is a gradual process, which takes time to work out, but comparison of the respective shares of the population of Somerset living in towns grouped according to size-categories between 1851, at the beginning of the railway age, and 1951, strikingly exemplifies this tendency towards concentration.

	% total county population	
	1851	1931
(a) 5 major towns : Bath, Taunton, Bridgwater, Yeovil and Frome	13	30
(b) next 10 towns	5	10
(c) 17 other towns (prosperous until mid-nineteenth century)	4	$4\frac{1}{2}$

The relative as well as the absolute importance of the larger centres has been enhanced at the expense of the smaller, which have receded into the rural background. The conformity of their population trend with that of the remaining rural population reflects the loss of their functional distinctiveness also.

To the extent that each phase of transport promotes a different degree of centralization it has its appropriate scale of urban mesh. The radius of the mediaeval market area, for example, was three to four miles, the distance that could reasonably be covered by a farmer at walking pace who had to drive his stock on foot to market and return on the same day. A statutory prohibition against setting up a new market within six and two-thirds miles of an established market reflects recognition of this scale. But railways allowed nodal market centres to extend their range to include areas much farther afield provided that they lay within walking distance of stations; and direct motor transport has further extended and in particular compacted the areas commanded by major centres, making the intermediate ones redundant. In this manner, the concentration of central services brought about by improved means of transport is reflected in a selective development of established towns. In the process those best equipped by reason of established size and developed functions, as well as by reason of situation, amass additional functions and services.

It is important to recognize the significance of size as well as natural setting in connection with this extension of nodality. Newcomers, e.g. industrial towns and resorts, that are introduced into the urban pattern compete as district centres with established service centres, and carve out fields of influence, though the force of their competition is not always commensurate with their size. Whatever may have been responsible in the first place for the aggregations of people that form towns, these by virtue of their size and the traffic they provide attract routes. The dictum of Vidal de la Blache, quoted above, may as truthfully be transposed to become "Les villes ont fait les routes". This is true not only of towns which are essentially products of commercial relations but even of other highly specialized towns. If a town, in spite of mediocre natural endowment as a route centre, succeeds in overcoming the attendant disabilities, in the process of its growth it acquires a measure of nodality that may be considerable. Its urban functions and size justify concentration of routes upon it. Such artificial nodality, very different in its basis from the natural nodality referred to earlier, is well illustrated by Sheffield. Sheffield lies at the convergence of a number of small Pennine valleys, but as route-ways these valleys until recently were only of purely local significance. They were not thoroughfares, but Pennine cul-de-sacs, framed by high moorlands. It was only because of the scale of industrial growth that the costly provision of railway connections through the watersheds by means of long tunnels was justified, so that the valleys became through routes. That the inherent disadvantages of the situation of the city, lying within the Pennine highland, and not, like Leeds, Nottingham and Manchester, just beyond its margins, have not been entirely offset is seen in the failure of Sheffield to attain comparable status as a regional centre.

Before we leave the subject of situation, its political aspects must be referred to. Among the factors that decide the fortune of towns none are more sudden and striking in their effects than political changes that radically alter the territorial frame of reference. This is especially evident in respect of towns that discharge administrative functions. These are related to areas of finite extent and are no more stable than the pattern of political geography. A sudden extension of the area of a centralized state gives a corresponding accession of importance to its capital. In the politically splintered Italy of the early nineteenth century, Rome was but a shadow of the ancient imperial city. In 1871, when it became the capital of a unified Italy, its population was only about a quarter of a million, but as central functions for the new state were concentrated in the capital, it grew rapidly and by 1931 its population had passed one million. Even a city such as Buenos Aires, which had been steadily growing as a colonial town and port during the previous period, experienced a sudden upsurge after becoming the capital of the Argentine Republic in 1880, and by 1909 had become one of the 'million' cities of the world. The growth of Paris faithfully reflected the territorial extension of the realm of the French kings. Each notable phase of expansion and administrative centralization has left its mark in great buildings and has been recorded in the extension and internal reorganization of the urban area.

State-building may invest hitherto insignificant or long moribund cities with capital functions that precipitate urban concentration. Prague, ancient capital of the kings of Bohemia, was launched upon a new career after 1918 as capital of the new Czecho-Slovak state, and grew accordingly. In contrast, the disruption of the Austrian empire left Vienna capital of a small and largely alpine state, an imperial capital deprived of its empire.

Capital cities are not alone in feeling the effects of changing political patterns. Ports are liable to find their hinterlands enlarged or truncated by such changes. Thus the events which left Vienna the outsize capital of a rump state, severed Trieste from the hinterland which the Austrian empire had developed for it in central Europe. The new political framework re-oriented trade in Europe in the 1920s. That of Bohemia, for example, was drawn to Hamburg, while Trieste was fenced into an extremity of Italy by a new international boundary, which left it only the barren Karst country as its indubitable hinterland. The new political frontiers of 1920 led immediately to the creation of Gdynia, near the old port of Danzig, and the maritime trade of the new state of Poland, a territory that had largely fallen within the 'natural' hinterland of Danzig, was henceforth transferred to Gdynia. Thus the port functions related to a physical situation at the Baltic outlet of the Vistula basin were transferred from an old urban site to a new one as a direct result of the drawing of a political frontier. Political considerations apart, Danzig could very well have continued to discharge these functions, and Gdynia was redundant. One more example will suffice. In Ireland, Londonderry has stagnated as a port and as a service centre since its trade area was seriously affected in 1922 by the creation of a new political frontier, which has cut off much of its former territory to the west and south-west.

Each of the great wars of the twentieth century has brought new frontiers which have led to profound re-adjustments in the territorial relations of towns, though the full effects of the present division of Europe, which has hardened along the Iron Curtain, have yet to be seen. Certainly they must affect the status as ports and towns of places such as Hamburg, Stettin and Königsberg (now in Russia and renamed Kaliningrad).

Distinct phases in the history of towns, corresponding to transformed situations, whatever the causes of these may be, are manifest in the external aspect or physiognomy of the urban area. Should a town enter upon a period of new activity and accelerated growth, the accession of new or enlarged functions is marked by the appearance of new organs to discharge them. Their introduction may necessitate considerable internal adjustments, perhaps especially evident in respect of the accommodation of space-needing railway and port installations. Old towns, such as Carlisle and Newcastle (Fig. 18) which have become modern railway centres, have experienced in the process considerable modification of their earlier lay-out to make room for the railways. The provision of satisfactory accommodation for bus stations has now become a town-planning problem that faces many cities.

On the other hand, the decisive change in fortune may betoken a check to growth and entry upon a period of stagnation or decay. Old functions are lost or depleted, new organs of urbanism fail to appear, and the active life shrivels within the old shell, which is more likely to persist than where continued growth brings with it a constant series of replacements and adjustments. One of the most perfect examples of a mediaeval urban shell is provided by Aigues Mortes in the Rhone delta, which enjoyed a short-lived prosperity as a port after the French kings acquired authority over the coast of Languedoc but before they came into possession of the Provençal coast, with Toulon and Marseilles. The town has never grown beyond the intact thirteenth-century ramparts.

It sometimes happens that a fossil town is infused with new life after a considerable interval through finding itself heir to a new situation. During the Middle Ages, Southampton was one of England's chief ports, a staple of the wool trade and chief link with the king's extensive French territories in Guyenne. Thereafter it suffered eclipse until the middle of the nineteenth century when, with the coming of the railway and the increased harbour needs of modern shipping, it has again risen to prominence as London's outport for liner traffic. Southampton still preserves clearly many features of its mediaeval core, including the gateway which miraculously survived aerial bombardment in the late war.

No phenomena illustrate the significance of changing geographical values more profoundly than the relative status of towns. True though it is that the genesis of a town is the outcome of the exercise by men of options with regard to conditions presented by their environment, equally the growth, maintenance or decline of its status are irrevocably bound up with the value of its physical setting under changing conditions of human culture.

Physical changes may be decisive for urban fortunes, as

c

when towns are destroyed and their sites abandoned as a result of the occurrence of natural catastrophes such as earthquakes or volcanic eruptions. Again, progressive secular changes, notably those associated with coastal evolution, may radically affect the bases of urban prosperity and even existence. These are special cases, however, and most towns are more or less immune from the effects of actual physical change because the tempo of most physical processes is much too slow to operate effectively during the few millennia which span the most permanent of town sites.

No town, however, is independent of the effects of changes in the cultural situation upon which the value of its physical setting depends. Cultural change must inevitably either strengthen or weaken the original value of the site and situation of each and every town. By creating new human needs and new capacities it is responsible for the emergence of fresh places as favourable localities for urban development, and for the obsolescence of others that were formerly auspicious. The weakening tendency, however, may be retarded or even largely overcome by human capacities of adaptation. Since the establishment and growth of the town have fixed capital, both material and social, at a particular place, diminution in the value of the physical setting is only exceptionally followed by evacuation. Human groups show extraordinary resources in overcoming growing handicaps to their continued occupation of an established site. Confronted with the alternative of moving or of striving to overcome mounting difficulties and disabilities a community almost invariably chooses the latter, and usually with a considerable measure of success. Yet when a survey of urban fortunes is extended over a long period of time, or its equivalent a period of accelerated or revolutionary cultural change, the distributional adjustments in the urban pattern are clearly apparent. The mill of geographical circumstance grinds surely, if sometimes slowly. Urban centres which have really profited from successive radical changes in human organization and technology, as distinct from those which have survived them, are few in number and deservedly great. These favoured few have proceeded from strength to strength, accumulating population, wealth and prestige.

So kaleidoscopic have been the revaluations of environment in Britain, America and many other countries during the last two hundred years that the contemporary urban scene is in the main an accumulation of residues, the deposits of a succession of cultural settings that no longer apply. The changes in urban status that have taken place demonstrate the appearance of new bases of urban prosperity, the strengthening and reinforcement of some old ones, but the undermining of others. No less striking, however, in view of the nature and scope of the changes, has been the demonstration of man's power over his environment through his capacity to adapt and accommodate old towns to new functions or new conditions. It follows, therefore, that the interpretation of the complex inter-relations of site and situation, form and function, which is involved in the geographical analysis of a town, can only be undertaken in the light of the sequence of stages in its development.

CHAPTER IV

TOWNS AND CULTURES

OUR earlier survey of the bases of urban development has
illustrated the variety of needs and circumstances responsible
for towns. The emergence of a town, no matter what its origin,
involves the creation of a distinctive physical and social
environment, which is 'urban', different from the countryside.
Yet the phenomenon has many and varied manifestations, both
in time and place, all embraced by the collective term 'town'.
Recognition of recurrent types of site, situation, or specialized
function may each serve as a basis of classification, and the
characteristics common to each different class find expression
in the urban geography. Moreover, by virtue of being the
areas of the earth most completely transformed and overlain by
man's erections, towns are among the foremost expressions of
different cultures. They are today almost world-wide in their
range of distribution, but they display a great variety of forms
that bear the stamp of different cultures. Some are very local,
others are widespread like the expanding, pervasive cultures
they express. National distinctiveness, too, finds one of its
most obvious indications in towns, so that an English town is
sensibly different from its French or American counterparts.
In each case, a specific national flavour is imparted by the
presence of distinctive social institutions in addition to the
basic urban ingredients, as well as by the forms the latter
take. Thus the public house in the British town, as the café in
the French town, have no complete parallels in the towns of
other countries. Along with a whole assemblage of other
features they express a way of life. They need not themselves
be exclusively or even primarily urban features, but they render
distinctive the towns of their particular cultural province. The
architectural prominence of religious buildings as urban landmarks throws emphasis upon contrasts in the faces of towns
that express differences of faith. The numerous nonconformist

chapels that are so conspicuous a feature of Welsh towns, no less than the mosques in towns of the Moslem world, are deeply expressive of the cultural tradition.

During the course of the nineteenth-century growth of our towns, mass production of housing did much to stereotype their residential areas. More recently, the process has been carried further by the development of multiple businesses in retail distribution and commercial entertainment, concerns that tend to operate over the whole national territory. Their adoption of distinctive styles of building makes the shopping centres of our towns conform increasingly to a standardized British type. The same range of shop and cinema fronts is displayed; a large proportion of the individual units are replicas of those that may be found in any other town centre in Britain. They differ only in their arrangement. But if British towns are thus becoming more and more alike, in the process their distinctiveness as compared with other countries is emphasized. It is on crossing the economic and social frontiers made by state boundaries that the forms change and we meet new styles of housing and shops. And, to a degree that we who are members of the particular society appreciate perhaps only faintly, the contemporary British way of life is epitomized in the semi-detached suburban villas with garden plots at back and front, which occupy great tracts of our towns. However noticeable and subtly significant the differences between towns in the countries of Western civilization, those of the East are still more different. Here the townscape mirrors a different world, and this notwithstanding the common and important function of the town as a principal medium of westernization, and perhaps its strongest expression in the oriental landscape.

Inevitably the characteristic institutions and habits of a society imbue towns with a distinctiveness which extends beyond architectural idioms, however striking these may be. It is an aura that pervades the whole town, and the family resemblance among towns that belong to a common culture is revealed to all the senses. It is not only apparent to the eye in the visible scene, but also to nose and ear through distinctive smells and sounds. Nor is the distinctiveness evident only within the town; it is obvious from a distance through its expression to a peculiar degree in the urban profile. The general character and major landmarks of urban skylines to some extent reflect different functional emphasis, so that there are appreciable differences between specialized types of towns such as manufacturing, port and fortress towns.

Even more radically do they differ from one culture type to another. The organlike pile of skyscrapers that mark the cores of American cities, and their precipitous descent to the low roof-plains of the residential areas, tiered as between tracts of apartment dwellings and of suburban villas and shacks, contrast with the more graded profiles of most west European cities, from which the real skyscraper is absent. The flattish profiles of Eastern cities throw into relief Americanized core areas where skyscrapers have recently intruded. Differences in habits of domicile, such as have been noted above, as well as in the distribution of historic architectural styles, which persist more especially in the monumental buildings, contribute significant and often striking features to the profiles of the towns with which they are associated. So much so, that characterization of urban profiles seems to deserve much more attention than it has yet been given. It is a highly significant aspect of the geographical description of towns.

A wealth of pictorial records and of literary descriptions supported by a few analytical studies, suggests that the recognition, characterization and distribution analysis of regional types of towns offer a promising field of geographical study awaiting systematic treatment. It requires more than the empirical study of contemporary towns. In this aspect of urban geography, as in others, historical depth needs to be brought to the survey. In so doing, not only is the appreciation of present differences enriched, new vistas are opened for study. As much as any other tracts, if not even more, towns present palimpsests where the past activities of man are recorded in the contemporary scene. In successive phases of human history and successive scenes of urban development, the town appears in new or different guise as the mirror of its age and the epitome of its region. Its equipment of institutions and the structural forms in which they are housed throw significant light upon the ideas and social habits of its people.

The great artificial platforms or *ziggurats* on which stood temples dominated the silhouettes of the ancient Sumerian cities, and reflected the role of the priestly caste in the governance of those earliest urban societies. Equally impressive and indicative of the government of another vanished society were the royal palaces of the Minoan cities at Knossos and Phaistos. The glorious Gothic cathedrals that grace many of the cities of northern Europe which date back to the Middle Ages bear faithful and lasting witness to the role of the Church in the life of mediaeval Europe. Nor could any clearer reminder of the political and social structure of the old Germany before Bismarck be invoked than its fantastic urban manifestations, the numerous 'court' towns of petty princelings. Each displays in capricious and whimsical details the fashions of a bygone age as they were adopted or created by autocratic rulers. Examples could be multiplied, but it will suffice to add mention of the public buildings, including forum, amphitheatre, and baths, which with pagan temples (*basilica*) were features of towns throughout the Roman Empire. Even in their ruins they conjure for us vivid pictures of the daily life of Roman citizens.

The features to which we have just been drawing attention pertained to towns of particular periods, and support the distinctions made by the historians of urbanism when they write of the classical, mediaeval or baroque city. But they were not universal features of their contemporary scene. As products of distinctive cultures, they invariably had a geographical frame of reference as well as an historical context. Urban geography is therefore concerned with recognition of significant groupings of such cultural traits as they represent regional types and with the establishment and interpretation of the facts of their distribution.

Professor Fleure has recognized a well-defined type of town in northern France, and has noted the features which are there very commonly associated in a fairly constant pattern. Many cities in the Seine and Loire basins date back to Roman times and their main streets may preserve the plan of a Roman camp that was typically sited alongside a river at a favourable crossing-place. Once the Christian religion had been established

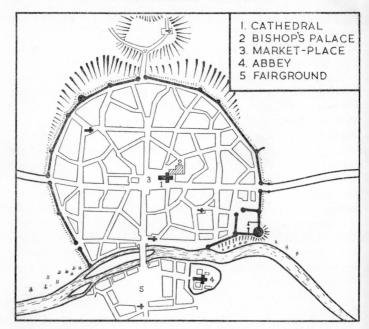

Fig. 9. Generalized diagram of a typical city in northern France
(after Fleure)

these Roman foundations became seats of bishops. It was
chiefly as such that they survived through the Dark Ages, so
that when urban life revived in the eleventh and twelfth
centuries, it often gathered round the bishop and his church.
The cathedral, with bishop's palace close by, thus became the
focus of the city's life and was usually situated at its heart,
alongside the market-place, and often at the crossing of the
main streets of the one-time Roman camp. The mediaeval
town was the abode of craftsmen and the centre of small industries catering for the needs of the surrounding population,
who came in on market and fair days. Trades were grouped in
streets named accordingly or from the churches with which the
trade gilds had special associations. On the outskirts of the
towns, outside the old walls, there may be a church that was

once a Benedictine abbey and an old fairground with a church nearby that is dedicated to St. Giles. A castle may be present but it is seldom the main feature, whereas the number of instances where the church is dominant is legion.

In Britain, on the other hand, continuity with Roman foundations is more rare, and the Church did not play the lead so directly in the resuscitation of town life after the Dark Ages. The cathedral, more usually an abbey foundation, is less typically contiguous with the market-place at the town focus. There are, however, many more examples of towns developed around and under the aegis of castles, the seats of Norman lords which sometimes replaced earlier and simpler fortifications. In the Paris Basin, again, few towns have mediaeval buildings expressing the corporate life of the bourgeois, though these developed impressively in regions farther north and north-east. In Flanders and the Rhineland historic town-halls and gild-halls are an outstanding feature of old towns, whereas the cathedral is rarely so central as in the Paris Basin.

The generalized diagram opposite incorporates the recurrent elements and disposition that are so typical of the cities of the Paris Basin. They differ in significant respects from the cities of other parts of western Europe, and even of the southern half of France, the Midi, country of the *langue d'oc* and a different cultural tradition. Here continuity of urban life back to Roman times is much more general, and more substantial survivals of Roman structures have had a correspondingly greater effect upon the development of the town plans. Features such as Roman amphitheatre and theatre formed subsidiary foci, so that the town-plan developed as a polyaster, and the mediaeval church was seldom able to dominate the scene alongside the central square, but was forced to fit into the plan as best it could. In so far as the territories grouped in modern nation states have shared a common experience this is reflected in common features of their urban geography; but in fact, as is well known, there are important differences within the national territories, as well as close affinities that transcend political frontiers, and both find significant urban expression. Much study of towns along these lines remains to be done. For

C*

European cities detailed analysis of regional types and plotting of their distributions will be illumined by the suggestive writings of Professor Fleure, and the conception obviously has a much wider application.

The epitomization of a distinctive regional history in a type of town is further emphasized by recurrence of a type of physical setting, as well as by repetition of a functional basis. The distinctive traits of the Hanseatic city, which have also been pointed out by Professor Fleure—the estuary-head position, the defensive water-girdle, the *Rathaus* dominating the centre of the town, with the mansions of merchant princes a conspicuous and notable architectural feature, rich in ornamental gables —show just such an intermingling of physical and cultural elements in the urban characterization. The dominant culture forms show a predilection for certain physical settings, so that the two become inseparably associated features of the town type. Much of the distinctiveness of the Mediterranean town is attributable to such a combination of a common experience and a similar physical setting.

Some of these town types are as highly localized as they are distinctive. They are the products of peculiar regional histories in which certain characteristics and episodes have decisively affected urban development. In the plain of Hungary there are the extraordinary 'peasant cities', e.g. Szegedin, Debreczin, Kecskemet, and Hodmezovasarhely, set amid a countryside that was quite recently empty of the usual forms of rural settlement. Some, though not all, of these extensive concentrations show a distinct urban nucleus where commercial and other services are grouped but the great mass of the settlement is simply an exceptional agglutination of the dwellings of agriculturists. They date from the Turkish invasions of the sixteenth century, when the countryside was depopulated and its peasantry gathered for protection in large pseudo-towns, which in character were outsize villages. The territory of many abandoned rural communes came into the possession of these towns. When their communal land was later parcelled out and re-developed the scattering of population that might have been expected to accompany the redistribution of land was delayed by ordinances designed to keep the citizens in the towns in order that they might the better respect their religious and civic duties.

Other town-types show a widely scattered distribution. Their close family likenesses are clear evidence of the dissemination of cultures. Urban forms have been carried far afield and transplanted into strange environments in colonial towns. Although modified in adaptation to their distinctive functions and settings their parentage is unmistakable. The respective motherlands are hardly less clearly recognizable in the colonial towns than at home. Batavia is as definitely Dutch as Singapore, Nairobi and Rangoon are British, or Saigon is French. Although a superstructure of American or cosmopolitan character has developed at its heart, Buenos Aires both in plan and fabric reveals its essential character as a Spanish colonial town, a type well represented in the New World. In the intermontane basins of the Carpathians, the stamp of the old Germanic city is borne by the towns whose foundations date back to the late mediaeval German colonization. Such are Sibiu (Hermannstadt) and Brasov (Kronstadt), in spite of their new names.

Towns are extremely potent agents in the spread of cultures and are among the most notable illustrations of the process whereby forms and patterns developed in a particular setting are introduced elsewhere as exotic features. Indeed, it is hardly an exaggeration to regard the town as the normal settlement form of imperial expansion and intrusion of civilized people into territories already occupied by societies engaged in practising a less advanced economy. The incomers find security in concentration and collect in towns for their mutual protection. Their urban character further expresses their role as administrators, traders or military pensioners, as compared with the indigenous rural population. Two of the commonest types of gathering point for town development, the fortress and the trading post, are very typical features of such an impact of cultures.

The numerous towns planted in the Middle East immediately after the conquests of Alexander the Great, and those later established by Rome as successive provinces were added to the Empire, were the chief agents of Hellenistic and Roman cultural extension. In the eastward spread of German culture during the early Middle Ages the foundation of towns played a no less important role; they were at once military strongholds, trading outposts, and centres of missionary enterprise among the heathen Slavs. Likewise, successive stages of Russian penetration into the territories beyond the Urals were marked by the establishment of incipient towns in the form of Cossack outposts and trading stations. Here, however, the agricultural colonization of these vast and almost empty territories, which belongs to an appreciably later phase, after the building of the Trans-Siberian Railway, was essential for the full assumption and realization of urban functions by these earlier foundations.

The way in which in Wales the towns, introduced by English conquerors and without roots in the native tradition, have been the channels through which English speech and ways have infiltrated into the Welsh countryside, is quite typical of the role of towns as transmitters of alien culture. It is noteworthy that in contrast with the essentially Welsh place-names of the surrounding rural areas, the towns in Wales have English, rather than Welsh, street-names.

Where the rural character of society is as deep-rooted and persistent as in Wales and Ireland towns may long exist as alien forms. That acute reporter upon Welsh society, Giraldus, noted that the Welsh had no towns, and in mediaeval Wales towns were symbols of alien domination. Small trading communities, incipient towns equipped with market-rights, attached themselves to the Norman castles established by the Marcher lords in their territories. Edward I's conquest of the mountainous and remote territory of Gwynedd in the northwest, last fastness of Welsh independence, was accompanied by his planting of garrison towns, as at Flint, Conway, Beaumaris, Caernarvon and Aberystwyth. In these planned and regularly laid out 'bastide' towns the king's henchmen lived under the protection of strong fortifications. From these strong points the Welsh countryside was held in subjection. For long the Welsh were debarred from living in these Anglo-Norman towns, and they had none of their own. The restriction of commercial transactions to the baronial markets was rigidly enforced.

Ireland, too, lacked any native urban tradition. It is admissible to regard the coastal bases established by Norsemen at Dublin, Waterford, Wexford, Cork and Limerick as the earliest germs of urban development there, but the continuous history of urban life in Ireland dates from the Anglo-Norman conquest. As in Wales, urban communities appeared under the protection of military bases, established at points on the east coast such as Carrickfergus, Downpatrick and Dublin. Outside them, urban ways of life made little impression upon the native Irish society, and even at the beginning of the seventeenth century Sir John Davies could remark upon the relative absence of settled life, still more of any urban habit, among the Irish.

The Jacobean plantation of Ulster, a planned colonization with settlers brought over from Scotland and England to the confiscated territories, was accompanied by a spate of urban foundations. It is interesting to note that in a memorandum Bacon presented to the king in 1609, an urban form of settlement was advocated for the newcomers. Both on military and economic grounds he strongly urged their settlement in fortified concentrations in preference to scattering them over the countryside. This would have proved very inconvenient for the exploitation of the farm-holdings into which the escheated lands were parcelled and it was not carried out. The influx of English and Scottish settlers was a colonization of farmers. It is not surprising then that among the many new towns envisaged by the comprehensive regional plan of 1609, and to which charters of incorporation were granted, some proved abortive. Others, however, did achieve fruition, and Londonderry and Coleraine were notable products of this essay in town-planning. Their rectilineal lay-out, their architecture in stone or half-timber, with gabled roofs of tile or slate, and even to some extent the actual building materials, were importations. Besides the town building that actually emerged from the plan of 1609, the private enterprise of alien landlords was responsible for creating during the course of the seventeenth and eighteenth centuries several new towns, designed as market centres. Towns were conceived and planted by the foreign aristocracy; no more than in Wales did they spring naturally from the structure of the native society.

The distinctiveness of the Irish town, especially in Ulster, owes much to the frequency with which it is a planned creation. The street pattern is often rectangular, and a typical if not ubiquitous feature is the 'diamond' or open central square provided as a market-place and focus. A central site was provided for the established church, but it reflects the long persistence of penal laws that Roman Catholic chapels are found typically only on the outskirts of Irish towns. Although the regular lay-out of the plantation towns persisted, few of their buildings survived the troublous seventeenth century, and the present character of these and many other Irish towns owes much to the extensive building that accompanied the greater political stability and economic prosperity of the late eighteenth century. It was fortunate that this expansionist phase coincided with a period of architectural distinction and good taste in Britain, so that Georgian urbanity is especially represented in the Irish townscape.

Urban development in Wales and Ireland contrasts strongly with that in the English lowland or in the Paris Basin, where towns are much more an autochthonous feature, springing from the life of the countryside and thoroughly integrated with it. Tensions between town and country are heightened where the geographical distinction corresponds to a deep social cleavage, the outcome, at least in part, of an intrusive culture. This is so often the case in east central Europe, where the towns are scattered islands of people of distinctive language, religion and ways of life. In Transylvania, Germans and Magyars, as well as Jews, are urban people amidst the Rumanian peasantry. Here, as in most plural societies, the urban-rural ratios of the constituent groups differ widely, so that the dichotomy of town and country usually has a special significance as a culture divide.

The impact of cultures, however, does not only find its typical expression in a disjuncture between town and country. It is frequently reflected in the internal structure of the towns themselves. Although the immigrants may be essentially town-dwellers, the indigenous peoples rarely remain confined to the country. They are recruited to the towns to perform menial tasks in the service of their urban masters, or are simply drawn there by the new opportunities for employment in the varied urban activities. The cleavage between groups is then apparent in geographical segregation. An interesting illustration is provided by old maps of Carrickfergus. This town, the oldest urban foundation in Ulster, was established alongside its Norman castle on the north shore of Belfast Lough early in the thirteenth century, and was one of the chief bases of Anglo-Norman penetration into Ireland. Along with Newry it survived as a foothold of English influence in Ireland, and became the natural headquarters of English power during the Elizabethan subjugation of Ulster and the plantation that followed. Many settlers came over to Carrickfergus at the beginning of the seventeenth century, especially from Devonshire, and built houses of English style,

"with brick and lime or stone and lime, well-tiled and slated, with handsome lights well-glazed"

beside the Lord-Deputy's mansion, Joymount. Contemporary maps show these substantial dwellings, but in other parts of the walled town there is a higgledy-piggledy, streetless sprawl of beehive-shaped huts, presumably the homes of the native Irish. Somewhat later in the seventeenth century distinct quarters developed outside the gates. To the north-east along the foreshore a colony of fishermen, brought over from Argyll and Galloway about 1665, formed Scotch Quarter, and outside the west gate there appeared a group of Irish, living in wattle and thatch cabins. This Irish annexe was probably a result of the 1678 proclamation which ordered all Roman Catholics to remove outside the walls of corporate towns. Scotch Quarter and Irish Town were names that long persisted.

In Spain the long period of Moorish occupation and the reconquest of the country by Christian kings have left their marks upon the old towns, which even in their present-day morphology show clear evidence of the succession of plural societies that occupied them. Although some towns, such as

Saragossa and Valencia, have a continuous history dating back to Roman times, four centuries of Moslem rule inevitably left modifications and additions. Apart from its mosques and enclosed bazaars, narrow, tortuous streets and numerous culs-de-sac were especially characteristic of the Moslem town; within the same walls, but otherwise separate, were the Christian quarter or Mozarabe, and the Jewish quarter. After the reconquest the Jews remained in their established quarter (Juderia), but the Moors were evicted and the towns were re-colonized with Christians, though not always without difficulty. Numerous churches replaced the mosques, and artisan colonists were established in districts (*barrios*) according to their places of origin and their trades, each gild being associated with a particular church. This feature persisted at Valencia and elsewhere until the eighteenth century. The Moors did not leave the urban areas altogether; they were allowed to occupy new quarters (Morerias) outside the walls, which were often protected by earthworks. By the sixteenth century, however, the Morerias had lost their identity.

The mediaeval towns of eastern Germany also showed distinctive German and Slav quarters, and the feature is equally typical of the urban products of modern imperial expansion. The European settlements in Africa and the East stand aloof from the native urban populations for reasons of health as well as of policy. In some cases, the present-day towns are essentially new concentrations around nuclei established by the incomers; in others, the agents of the colonizing power have superimposed themselves upon older native settlements. Dakar and Kano respectively exemplify the two processes; but in both the European and African towns are clearly separated by a *cordon sanitaire*. As in the East, the European town contains a population more numerous than the few European traders and administrators who live in a district quarter within it. At Kano in northern Nigeria the British 'township', outside the walled native city, included in 1937 only 360 Europeans in its population of 7,000. In this 'township', where much land is reserved for, or occupied by British administration, the 7,000 inhabitants are spread over nearly as great an area as is enclosed within the walls, where more than

80,000 Africans live. Fulani and Hausa, as well as other immigrant peoples, dwell apart as separate communities there, but in its external appearance the African city differs little from one ethnic quarter to another.

The European concessions in Shanghai and other Chinese cities, where extra-territorial rights have been enjoyed by the European powers, are mainly occupied by native workers. In Indian towns the separate quarters established by the British during their rule, though mainly occupied by their employees and dependants, were none the less distinct and deliberately separate from the indigenous towns. Spate has described the juxtaposition but clear separation of the

> "overcrowded and irregular Indian city of trades and handicrafts, and the more spacious and often mechanically laid out and anglicized 'civil lines' and 'cantonments'."

The 'civil lines' contain the official residences of the local bureaucracy, the 'cantonments' are the barrack quarters.

Since the establishment of the French protectorate in Morocco almost all the towns have been doubled. Marshal Lyautey set up French towns alongside, but rigorously separate from the indigenous towns such as Fez. The native towns of North Africa are often themselves clearly subdivided between Moslem and Jewish quarters, known respectively as the Medinah and the Mellah. Two thousand years earlier, the cosmopolitan city of Alexandria was clearly demarcated into royal or Greek, Egyptian, and Jewish quarters. In many parts of Africa large-scale migrations of detribalized natives recruited into industrial employment in the growing towns have produced on their outskirts the appalling shanty-towns or *bidonvilles* that have become so characteristic a feature of African urbanization. The squalid and revolting colonies of corrugated iron and hessian shacks which disfigure the western locations of Johannesburg are but particularly flagrant and notorious examples, emphasized by the speed and scale of industrialization, under rigorous conditions of 'apartheid'. Whether or not a relationship of political or cultural dominance and subordination is involved, communities of different origin, so long as they remain distinct in language or religion, tend naturally to be segregated for residence and social institutions. In Canada, where French and English-speaking communities live together on equal footing, Montreal, in effect, consists of three cities that are fused together rather than integrated. Besides the French and English, there is the Jewish city, and all major institutions are triplicated.

In these ways, the internal structure of towns faithfully reflects the impact and relations of distinct communities. But towns are also the natural channels by which extraneous elements are introduced into the body of society. For at least some towns are really cosmopolitan, points of contact between cultures. Aliens are present in incompletely assimilated groups, contributing still further to internal differentiation of the urban area. In the Greek city, trade was carried on by foreign merchants (*metics*) not by the citizens, and colonies of foreign traders are always a feature of commercial towns.[1] Those of the past have left many traces in urban place-names, indicating where they lived or carried on business. Thus Lombard Streets are a feature of several old-established cities of northern Europe, including London. Wherever a particular commercial link brought distinctive groups of foreign traders or craftsmen, distinctive quarters appeared. Sometimes, as in the case of the Huguenots, the immigrants sought religious or political asylum.

Most remarkable and ubiquitous of all such alien communities in towns through the ages have been the Jews, preserving their identity through all vicissitudes. The bonds of their own group life, and the ostracism to which they were subjected by citizens of other faiths, kept them together in distinct Jewish quarters. This feature developed to an extreme degree in the cities of eastern Europe as the ghetto. Jewish colonies are still the most widespread examples of ethnic quarters in towns, but distinct 'coloured' quarters in the dock areas of the great ports of the Western world are also a familiar feature, as for example Limehouse in London, Tiger Bay in Cardiff, and the Chinatowns of Liverpool, San

---

[1] In Constantinople, for example, the Venetians, Genoese and Pisans, who are estimated to have numbered sixty thousand in 1180, were settled in distinct quarters along the Golden Horn, each with its quay and church.

Francisco and other ports. Equally well known are the coloured quarters, such as Harlem in New York, which have become established in the northern cities of the U.S.A. as a result of immigration of negro labourers from the rural South. In Chicago, restrictive covenants, real estate agreements and other practices have confined the negro immigrants who have swarmed in since the first World War to a district of once-luxurious, but now dilapidated, mansions near the lake-shore. Central South Side, the select abode of the wealthiest classes in the late nineteenth century, has now become, in its progressive deterioration, the negro quarter, its great houses divided repeatedly into smaller and smaller apartment units.

The social geography of the contemporary American city only partly exhibits the various ingredients that have so recently contributed to the population. The immigrants belonging to the more recent streams congregated in compact groups to form ethnic quarters. These are still represented, but most of the urban area today is socially differentiated on the typical American basis of income-groups, irrespective of the antecedents of the inhabitants. Thus while there may be a clearly recognizable Italian quarter, Welsh and Irish quarters have disappeared, though many Jones and O'Reilly families are scattered about the town among other Americans. Dispersion of European immigrants from an ethnic quarter marks a significant stage in their assimilation. At this point they have become American.

CHAPTER V

THE MORPHOLOGY OF TOWNS

I. URBAN REGIONS

IN preceding chapters emphasis has been laid upon the study of towns as features which contribute to the distinctiveness of country, imparting individuality through their numbers, their arrangement, and their variant types which are among the clearest expressions of culture differences. But towns are themselves areas of appreciable size, and have an internal geography that is full of interest and significance. When, within any urban area, we recognize industrial belts, shopping areas, residential quarters and suchlike, we are expressing the internal structure of the town in terms of different users of urban land. It can also be described in terms of the physical forms and arrangement of the spaces and buildings that compose the urban landscape, or townscape, as it may be called. Differences in either or both these intimately related aspects of urban morphology, function and form, give a basis for the recognition of urban regions. It is the description of their nature, their relative disposition, and their social interdependence that constitutes a geographical analysis of an urban area.

As an illustration let us examine St. Albans. The Abbey, founded at the end of the eighth century, was the pre-urban nucleus under the stimulus of which the old town developed. It became a busy coaching centre in the early nineteenth century. The later growth of the town was speeded up after the coming of the railways, and has been accompanied by increasing internal differentiation. This has recently been the object of intensive study by Mr. Stewart Thurston, whose detailed survey of current land-use, interpreted by analysis of the historical records, especially in the form of maps, will be drawn upon to illustrate this aspect of the geographical study of towns.

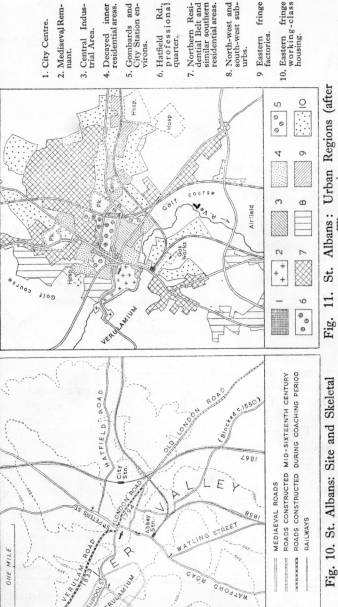

1. City Centre.
2. Mediaeval Remnant.
3. Central Industrial Area.
4. Decayed inner residential areas.
5. Gombards and City Station environs.
6. Hatfield Rd., professional quarter.
7. Northern Residential Belt and similar southern residential areas.
8. North-west and south-west suburbs.
9. Eastern fringe factories.
10. Eastern fringe working-class housing.

Fig. 11. St. Albans: Urban Regions (after Thurston)

MEDIAEVAL ROADS
ROADS CONSTRUCTED MID-SIXTEENTH CENTURY
ROADS CONSTRUCTED DURING COACHING PERIOD
RAILWAYS

Fig. 10. St. Albans: Site and Skeletal Plan

Surviving as an enclave near the heart of the modern town is a mediaeval remnant, containing the Abbey and its precincts. It has been by-passed since 1833 when a new coaching road (Verulam Road) replaced Fishpool Street as the town's western outlet. In this remnant of old St. Albans houses are interspersed with inns and tiny shops, the buildings forming a continuous closed frontage along the winding street. Very few of them are modern, and the only substantial modification they have experienced has been the addition of false Georgian fronts to a few of the larger old houses.

While this part of the old town has remained intact, the area immediately to the north-east, the City Centre, has been subjected, directly or indirectly, to every new influence that has impinged upon the town. Shops, offices, banks, public services, as well as the market, are grouped here at and near the principal route-focus. As the town has grown and acquired new and enlarged functions, this core has been greatly changed in its fabric, if not in its street-pattern. Since 1918 its invasion of the adjacent area to the north has driven a wedge along St. Peter's Street into a formerly residential area, and the older buildings have either been replaced or altered to fit them for new uses. Municipal administration, which has now left the Town Hall, and other offices of central and local government have found accommodation on the east side of St. Peter's Street in old buildings that have been restored.

The City Centre is flanked to the west, but more especially on its east side towards the railway, by the components of what Thurston has termed the Central Industrial Area. A product of the railway period since 1860, its extension as an industrial area had become stabilized by 1914, though its characteristics have undergone some subsequent modification. Factories, especially for printing and clothing industries, sprang up here within and on the outskirts of the growing nineteenth-century town. They were small, congested, and closely intermixed with small, mean houses. They had to grow upwards rather than outwards, and their many-storeyed buildings, crowded into cramped yards, are very typical of their period. Now out of date, they change hands frequently, and tend to be subdivided among a number of smaller, miscellaneous undertakings. Along the margins of the City Centre some of them are now used only for storage, and the impression of an incipient warehouse zone is further emphasized by the appearance of a few newly constructed warehouses. Other signs of decay are the many small workshops, of cycle-repairers, cobblers, joiners, and builders' merchants, which occupy all or part of old houses, their yards and outhouses. The densely packed, tiny nineteenth-century dwellings of this area still house six or seven thousand of the poorest classes, whose day-to-day shopping needs have produced another feature, likewise indicative of declining status, the small street-corner general store, which usually involved little more than fitting counters and shelves, and perhaps a larger window, in the front parlour of a house. The ground-floors of the larger villas that flanked the main thoroughfares (Victoria Street and London Road) as they traverse this area have also often been converted into small private shops.

Terraces of tiny, tunnel-back dwellings appeared during the latter nineteenth century on the few vacant plots in the Central Industrial Area. Large compact blocks of similar terrace-houses were built in a broken fringe of estates on the outskirts of the growing town. These do not show the extreme congestion that exists in the centre, but most of the property falls far below modern standards of housing, and these estates have also acquired those symbols of residential deterioration, the street-corner store and the small workshop. Some local service industries, such as bakeries and laundries, have infiltrated, but not other factories.

Even before 1914 overcrowding in the Central Industrial Area had caused several of the larger firms to migrate to the outskirts. They needed railway frontage as well as space, and the considerable lengths of the railway approaches to St. Albans that are confined within cuttings meant that this new factory development was largely directed to the open country east of the main railway, along the Hatfield branch line. Here large plots were available and single- or two-storeyed factories were set up. Since 1918, further rapid extension of this factory area has proceeded, chiefly for clothing and light engineering industries. In its present character, however, the area has

become essentially dependent upon road transport, and few
of its factories now make much use of railway sidings. To house
the growing numbers employed in this industrial tract, as well
as in Hatfield farther east, considerable recent extension of
building has taken place on the east side of the town, filling in
and engulfing the earlier incomplete girdle of housing. A broken
arc of modern estates of small villas has appeared. They contain
a few subsidiary shop clusters, but the dominant impression is
of dreary monotony of houses, standardized in styles and
regimented on their sites. Yet, although they are more crowded
and shabbier than the better residential areas, they compare
favourably with their nineteenth-century counterparts.

Besides stimulating industrial development, the railways
opened up St. Albans as a residential town for those whose
work lay elsewhere, including London. Around the City
Station substantial villas sprang up in spacious plots. This small,
but very distinctive area, the City Station environs, has not
since changed much in appearance, but its residential character
has been altered. Many of the three-storeyed houses are now
divided into flats and a few have been converted into private
hotels.

The increasing middle-class population of tradesmen,
office-workers, and industrial managers was largely accommodated in the nineteenth-century extension of the northern
margin of the town. This residential belt has since been invaded
by the expanding commercial core, which has breached it
along St. Peter's Street. To the west, the small residential
'island' of Gombards has been left, and is now rather similar
in character to the City Station environs. Its nineteenth-century
villas, many of them three-storeyed and set in wide, pleasant,
tree-lined streets, have retained a considerable measure of
their residential status, although many have been converted into
flats. East of St. Peter's Street, considerable adaptation of the
older mansions for professional purposes bids fair to create a
professional quarter along Hatfield Road. Here most of St.
Albans' doctors and nursing homes are found; and some
central government departments, as well as the Rural District
Council offices, have established themselves in the large
houses. The Loreto School and the County Museum add

variety to a district which has experienced no great change in appearance since it was built.

The newest, northernmost part of the residential area as it existed before the first World War has since been added to. Whether detached or semi-detached, all the houses in this northern belt of middle-class residences have a generous street-frontage and sizeable gardens. The impression of spaciousness is enhanced by the presence of tennis-courts and bowling-greens. Shops are notably absent. Other middle-class residential areas, to the south and south-east, have spread out along Watford Road and London Road. Their older parts, consisting of pre-1914 mansions straggled in ribbon development, have sometimes experienced functional change like that along Hatfield Road.

Finally, on the outskirts, especially to the north-west and south, lie widely separated areas of open suburban development in the form of small estates and individual buildings— a haphazard intermixture of styles and classes of dwelling, interspersed with market-gardens and other open spaces.

A single example such as we have just been considering cannot be expected to provide complete illustration of the manifold combinations of features that make for the internal differentiation of urban areas. Yet even St. Albans, a modest town of fewer than 50,000 inhabitants and one with a highly individual history, demonstrates many principles that have wide application.

In the course of urban development, certain forms of land-use and building become segregated while others are brought together in close association. Recognition of the spontaneous nature of functional differentiation as a process that is part and parcel of urban growth does not imply acceptance of a particular ontogeny such as has been postulated by Griffith Taylor in his description of the seven ages of towns. Using the terms infantile, juvenile, adolescent, early mature, mature, late mature, and senile, he describes successive stages of functional differentiation which are presumed to have general application. Without straining the argument so far, however, the cumulative effects of certain centripetal and centrifugal tendencies can be clearly discerned. Under pressure of increasing congestion in the centre, trade and central services, the functions most in need of central sites and best able to compete for them as land values rise, concentrate there, while residence and industry tend to move out. In so doing, their relative dispositions are affected by their different requirements of site and communications, as well as by their own mutual relations of attraction and repulsion.

Thus in St. Albans, modern factories, seeking open, flat land alongside transport facilities, have gathered on the east side of the town. Their proximity has been shunned by middle-class residences, but has attracted working-class housing for the convenient accommodation of their employees. Consequently, we find working-class estates especially on the east side, whereas other residential development has kept to the north, north-west and south. Residential segmentation is a familiar feature of urban geography, and soon makes its appearance as a town grows. Differences in density and style of housing point to the social classes for whom the property was built, while the age, degree of obsolescence, and mode of occupation of the buildings are reflected by their present occupants.

Sites on high ground, well drained, airy, and commanding extensive views are claimed for high-class residences which not only avoid less favourable terrain for building but also undesirable neighbours such as factories, railway stations and marshalling yards, and port areas. The upper and lower parts of a town are often socially upper and lower class as well. The incidence of atmospheric pollution also plays a part in these social distinctions. It might be going too far to suggest that it is ever decisive in determining that the distinction between fashionable and slum quarters should be between west and east ends, but it certainly emphasizes the advantages of the former in our climate of prevailing westerly winds.

Urban Zones

(a) THE CORE

It has been seen how in St. Albans a specialized core, the City Centre, has emerged. In course of time, it has swollen at the expense of the adjacent residential area, and it is now showing signs of internal specialization because the site requirements of the various functions it accommodates are not the same either in degree or in kind. Some, notably shops, seek display frontage along the busiest thoroughfares. For others, such as nursing homes, hotels and administrative offices, this is much less important, and some measure of seclusion may even be preferred provided convenience of access is not seriously sacrificed. These trends, apparent in St. Albans, are much more fully developed in some cities, especially the largest. The core becomes especially distinctive, and is recognized by the use of a special name. Although this may derive in the first place from a description of its position, as, for example, the West End of London or the Centro of Buenos Aires, or its shape, as, for example, the Loop of Chicago, or the Triangle of Sao Paulo, its connotation becomes increasingly identified with functional distinctiveness. It is often the City or Town in a special, narrower sense, and in current American usage it is the 'downtown' district. Its own internal differentiation into quarters is often highly developed. Finance and insurance, public administration, markets and wholesaling, shopping and entertainment all tend to occupy separate districts and individual streets often become so highly specialized that their names become synonymous with particular activities, e.g. Fleet Street, Harley Street, Savile Row, and Hatton Garden in London, or Wall Street and Broadway in New York.

The patterning of the various activities that are concentrated in the urban core expresses a fine degree of adjustment to a complex variety of considerations. Activities that share the same desiderata of site and accessibility, seeking the patronage of the same clientele or linked in other forms of ancillary relationship or interdependence, congregate. They depend or thrive upon being together. At the same time, they are dissociated from other groups of activities with which their co-existence may even be incompatible. The neighbourhood associations of Billingsgate and Bond Street are far from being haphazard.

(b) THE INTEGUMENTS

As the urban core develops, the surrounding tracts appear as integuments of differing character, the products of successive phases of urban growth and the accompanying functional changes. The growth of every town is a twin process of outward extension and internal reorganization. Each phase adds new fabric—outside in the form of accretions, within, in the form of replacements. Replacement of old buildings by new ones, specially constructed for new functions and current standards, is always in process but is never complete, so that internal changes always include much adaptation of old forms to new uses. At any time, many of the existing structures are obsolescent and in their deterioration are subject to functional change; they are converted for new uses.

The inner and therefore older integuments immediately enclosing the central core are specially subject to change of function. They represent the residential and sometimes the industrial areas of earlier phases; a mixture of functions is now typical. In part, they have been invaded by extension of the business core and so absorbed into it. Elsewhere their obsolescent buildings on cramped sites have usually found new occupants. Intrusive features have appeared which are obvious signs of residential deterioration.

Much manufacturing industry, belonging to types tolerant of lack of space, may be accommodated in the crowded inner parts of the urban area in small factories and workshops. As at St. Albans they are in part residual industries that have resisted the centrifugal tendency to move out to new sites, but to a greater extent usually they are invaders that have colonized sites and even buildings vacated by earlier industries or residents. Typically, they occupy converted premises, old property that has become cheaply available in the side streets off the shopping thoroughfares, and in the still crowded but decayed residential zone surrounding the core. Here they exploit the advantages of accessibility for their employees and proximity to their clientele, which are often the shops nearby. Thus inner London, within the area of the metropolitan boroughs, has in its West End, especially in the area north of Oxford Street, one of Britain's major concentrations of various branches of clothing manufacture. The East End in Stepney has another, and also one of furniture making, in Shoreditch and Bethnal Green. Again, the fashion clothing (*haute couture*), jewellery and toilet preparations (*articles de Paris*), which are so special a feature of the industrial production of Paris are largely derived from the inner *arrondissements*.

The manufacturing activities of the inner areas of cities are especially associated with their assumption of a mixed functional character, small factories and workshops intermingling with dwellings, but units of particular industries may sometimes occur in such concentration as to form localized industrial quarters like the jewellery and gun quarters of inner Birmingham. These are relics of the workshop phase of Birmingham's industrial development, before 1860. Even before the end of the eighteenth century, the gun trade had become localized in the north of the town around St. Mary's Church, where it was carried on by small masters in workshops in or behind their own houses. There it survives, though its products have become increasingly specialized and are now mainly sporting guns. The jewellery trade, in the course of its rapid development during the first half of the nineteenth century, experienced a series of shifts, as during periods of expansion the manufacturers moved to large premises on newly built estates. By 1865, however, the jewellery quarter had become established north-west of the heart of the city, in the vicinity of Vyse Street and Warstone Lane, where it still forms a highly distinctive district.

In contrast with the confused intermingling of housing and industry which typifies the inner and older parts of St. Albans, adjoining the business centre, the later development of the town shows a clear differentiation of the two, even though blocks of working-class dwellings are closely associated with belts or blocks of factories. They are contiguous but not confused. This contrast in the relations of industry and housing in the inner and outer integuments of urban areas is very typical. Indeed, it may be claimed that, after differentiation of the central core, it is the most significant and widely represented feature of functional zoning. The core emerges in consequence of specialization upon non-industrial, non-

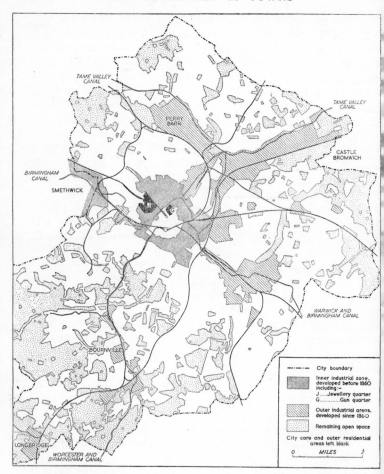

Fig. 12. City of Birmingham—core and integuments

residential functions such as commerce, administration and other central services. Surrounding it, the integuments are dominated by housing and industry. An inner zone is traversed and penetrated by radiating prongs of commercial development, but is mainly devoted to housing and industry. These

unctions are intermixed and occupy buildings that are often
ot clearly distinguished in their external appearance, since
heir present users have often adopted and adapted older
tructures. In the outer zone residence and industry are
ontiguous but separate, and the forms of the buildings they
ccupy are no less clearly distinct; shopping and central services
re present in the form of outlying clusters, localized amid the
racts of housing.

These fundamental contrasts between inner and outer
nteguments, evident even in St. Albans, are highly developed
n large metropolitan cities. Often they have been further
ccentuated by distinct changes in industrial emphasis since
he middle of the nineteenth century. In Birmingham, for
nstance, whereas an earlier phase of workshop industry was
oncerned with the gun, jewellery, button and brass trades,
ndustrial development since about 1860 has been along new
nes. Some of the older industries have moved out from
nner Birmingham to more spacious sites, as happened notably
vith the manufacture of cocoa and chocolate, which Cadbury
ransplanted in 1879 to Bournville; but more especially they
ave been new industries—cycles, electrical apparatus, rubber,
notor vehicles—carried on in large factories, that have devel-
ped on the outskirts of the growing city. They have tended to
oncentrate in tongues along the radiating canals and railways
nd where suitable land was cheaply available, as on the heaths
nd in the floodplains of the Rea and later the Tame. On the
orth side of Birmingham, the Tame valley is now heavily
ndustrialized right from Perry Barr to Castle Bromwich
'he segments that lie between the factory tracts have been
lled in with housing. During the nineteenth century, this
ypically took the form of monotonous and regular terrace-
ouses, in serried ranks, but since the first World War a much
nore open development has swallowed semi-rural villages four
r five miles from the city centre. Essentially similar contrasts
etween the industrial character and pattern of inner and
uter integuments are evident in Greater London and Greater
aris, where the peripheral factory belts show pronounced
oncentrations along the arterial communications.

The Population Structure of Cities

The emergence of a core, surrounded by integuments, which is especially evident in great cities, finds strong demographic expression. As the seat of commerce and other highly centralized services, the core is the hub of the city's life by day but is largely deserted by night, although part of it may form a 'bright-light' district in the evening and far into the night. Since it is the overnight population which is enumerated by censuses, displacement of residence by other functions creates an area which appears as an extending crater in the population structure. The commercial core, its tall buildings swarming with workers by day when its streets are choked with vehicular traffic and its pavements thronged by milling crowds, is at night abandoned to a small maintenance staff of 'caretakers and cats'. Even the bright-light areas, agog in the evenings after commercial and administrative quarters have closed down, empty later and are so recorded by the census.

So it is with all towns. Old towns that have experienced modern expansion have often suffered such a transformation at the centre that the resident population within the ambit of the old town has been steadily decreasing during most of the modern phase. In fact, once a certain stage is reached the population of a town can only be increased by extension of its boundaries. In great cities the population void at the centre spreads rapidly, as more and more wards or census enumeration districts pass their peak of population and become increasingly appropriated for non-residential purposes. Beyond the area that has already emptied the inner integuments, although still standing out as the most crowded areas, are losing population. People moving out from them join with immigrants to the town to colonize rapidly growing suburbs, where building is filling up remaining spaces and extending the outer margins of the urban area. The London of the late nineteenth century, defined by the new county that was carved out of the ancient counties in 1888, reached a population of four and a half millions in 1901, but has since been emptying into the outer zones of Greater London. Whereas the L.C.C. area now has fewer than three and a half million inhabitants, its overspill, together with immigration from outside, have increased the population of the outer zones of the conurbation from three to five millions in the last thirty years. The evacuation of the City of London, the square mile enclosed within the mediaeval boundaries, began much earlier, and from a peak of 128,000 in 1851 population has fallen to a mere 5,000. In part, these re-distributions are the natural outcome of the centripetal and centrifugal tendencies at work. But in part they are the direct result of social policies of surgical treatment of blighted areas by slum clearance and town-planning schemes, with the transference of displaced population to housing estates on the outskirts, such as the L.C.C. estates at Watling and Becontree. Birmingham has experienced uninterrupted growth since its incorporation in 1838 but is really only exceptional in having as an administrative unit more nearly kept pace with the physical extension of the built-up area. From a Civil Parish of 2,660 acres, containing 70,000 people in 1801, it has grown into the modern City and County Borough of more than 51,000 acres and well over one million inhabitants. In this process the central wards have had their densities drastically reduced since the '60s and '70s by the combined effects of appropriation of sites for business purposes and slum clearance, such as the famous Corporation Street scheme, associated with the name of Joseph Chamberlain.

Azonal Factors

The example of St. Albans has also shown how the outward extension of a town is typically uneven, unlike the annular growth of a tree-trunk. Building takes place most vigorously along the radiating arteries, while the intervening sectors are left to be filled in later. The arteries are also prongs from the core along which changes are transmitted. They exhibit a more advanced stage of functional succession and structural replacement than the areas behind, which their new facades screen from the view of most visitors. It often happens that a frontage that has been largely rebuilt or adapted for shops and business premises conceals a tract of slum dwellings.

Whereas in entering a city by any of its main roads one is met by an extending arm of the commercial core, the railway approach provides a very different impression. Although

D

factories and storage yards tend to line the railways, stretches of the unprepossessing backs of dwellings are also disclosed. Both road and rail arteries have their own forms of ribbon development to complicate the concentric zoning. The railway traverse, however, is usually the more revealing of the character of the integuments, except where the track is buried in tunnels or sunk in cuttings; but it is also often incomplete, since the termini as a rule lie outside the innermost core.

It is evident therefore that although there is a tendency to zoning, both of functions and of the forms that express these functions as well as the date of urban extension, the zoning must be imperfect. It is not to be expected that towns will show the idealized arrangement of a central core enclosed by successive complete zones. Certain very prevalent types of site, notably those of seaport or seaside resort, where the nucleus is attached to the coast, preclude this. At most the zoning here must be limited to a sector; but the internal structure will in all likelihood take shape in relation not merely to the initial nucleus attached to a limited stretch of water-front but also to lateral extension of this developed water-front itself. These are only more obvious examples of how the site introduces complications into zonal structure. Towns sited on the banks of important rivers are the rule rather than the exception, and rarely is the progressive extension of the urban area balanced between the opposite banks. Until the nineteenth century, London lay essentially on one side of the Thames; its vast extension since then has made south London the larger element both in area and population, but as is well known, the internal structure is by no means balanced as between north and south banks.

Established patterns of roads and railways and special conditions of site are controls which invariably introduce lack of symmetry in the growth of towns. It may be noted in passing that peculiarities in the former can in turn often be related to physical features, so that the influence of site is manifested partly through the close adjustment to it shown by the routes that converge upon the growing town, shaping both its external outline and its internal structure.

Let us turn again to St. Albans to see the effects of such

factors in its growth (Fig. 10). From its nucleus round the
Abbey, built where a platform of higher, firm ground abuts
on to the flat, damp, alluvial valley-floor of the river Ver
opposite the ruins of Roman Verulamium, St. Albans has
progressively spread over this platform between the two
rather steep-sided dry valleys that flank it. To the old-established lineaments of the road system which gave the town its
external contacts and which have shaped the developing plan
of streets, the coaching era added some important new lines—
London Road (1794) and Verulam Road (1833). Henceforth,
the axes controlling the chief extensions of building and the
infilling pattern were the newly aligned south-east to north-
west thoroughfare and that which intersected it at right angles
in the city centre. Roads approaching from the north and
north-east converged near St. Peter's Church or were directed
along St. Peter's Street to the market-place at the carfax
before making the crossing of the Ver valley. Beyond its flats,
once firm ground was regained on the opposite side at St.
Michael's Church, the road branched towards London (old
Watling Street) and Watford. The alignment of the railways
and the location of the railway stations have also been important in directing the later growth during the past century.

Spread of the town was for long confined by the Ver valley
to the platform on its north-east side. Only more recently
has building straggled along and spread out from the diverging
roads across the Ver valley, and in the south-west sector a gap
in the peripheral development remains. Here rural countryside
comes close to the city centre. The influence of the physical
nature of the Ver valley is here reinforced by the circumstances
of land-ownership in the area beyond it. For the Gorhamburd
estate and the considerable areas that have been Crown Lands
since the dissolution of the Abbey have never been made
available for building. Thus Verulamium remains an open
site.

St. Albans is in no degree exceptional in its illustration of
the importance not only of the physical background of site
conditions and the skeletal framework of communications but
also of the accidents of land-ownership in shaping the outward
growth and the internal structure of urban areas. There are

factors at work quite independent of a zoning principle, and even at variance with it, which predispose the form of extension and succession, making it easy and rapid in some directions and quarters, but retarding it or holding it up completely in others. It is very important to realize, however, that the various categories of urban land-users do not differ simply in terms of greater or less need of central sites. They do not fall so simply into a sequence of dispositions in relation to the centre. Their site requirements differ and terrain that is well suited for certain purposes is just the opposite for others. Broken hilly ground holds out definite attraction for residential development but is far less satisfactory for industry. Conversely, flat, low-lying land along alluvial valley-floors and even reclaimed foreshore is often very suitable for factories, especially as it is favoured in respect of rail and water transport. In some industries this is further emphasized by use of the waterbody for processing water or the discharge of effluent. These areas, however, are far from attractive for housing and their avoidance has often preserved them from building until modern times, when they have offered spacious tracts for transport and industry. Because of freedom from the competition of other types of user, such land, even in close proximity to areas of high land value, has been available for these purposes relatively cheaply. Once these functions are established, however, the attendant noise, dirt, or smell causes better-class residence to forsake or avoid their vicinity; but closeness to factories or docks stimulates the provision of low-grade housing for the labourers employed locally. In respect both of site conditions and neighbourhood a principle of dissociation as well as one of association is thus at work in the patterning of urban areas.

Enclaves

The symmetry of schematic representations of town growth is often impaired by the presence of enclaves of various kinds. With its mediaeval remnant and open south-western quadrant St. Albans once again provides illustrations of the importance of such enclaves in the morphology of urban areas. In the case of the mediaeval remnant of St. Albans, vestigial features, residues surviving from a remote past, form an enclave preserved more or less intact against the impact of the successional change that is so evident in other parts.

Enclaves are, in effect, reservations for special functions and their ancillary services, set apart from the free interplay of forces that make for the normal patterning of a town. They are tracts of urban land preserved from the operation of a market economy which assigns land to the highest bidder. Where such enclaves accommodate functions which date back to pre-urban antecedents or to early phases of urban development, they are quite likely to be situated near what has now become the commercial core. Examples are the precincts of palace, castle, cathedral (especially if originally an abbey foundation) or university, and their separateness may even have been emphasized by walling.[1] In association with such enclaves, open spaces have often been maintained, for example as royal parks or cathedral closes, while other central areas have filled to the point of extreme congestion. By-passed by the main streams of circulation, they are islands of peace and seclusion compared with the maelstrom of traffic nearby.

As at St. Albans, circumstances of land-ownership have sometimes preserved wedges of open space, like Newcastle Town Moor (Fig. 13b) and Southampton Common, extending right to the heart of the modern city. They have been sterilized against building development throughout the phase of modern expansion. Even parks and cemeteries of relatively recent establishment, laid out during the nineteenth century on what were, at the time, the town outskirts, have often been engulfed to become enclaves which are now jealously protected as islands of open space in a sea of buildings. Albert Park, Middlesbrough (Fig. 20), established in 1868, at the height of Middlesbrough's growth, is but one example among a great many. Much better known is Central Park, New York, which helps to confine the city core in south Manhattan. Beyond it and on each side, the tide of building has spread over Manhattan Island and on to the mainland, where Van Cortland and Bronx Parks have in turn been enveloped.

[1] The most extreme type is the city within city, exemplified in Pekin, in the Moscow Kremlin, or in the Vatican City of Rome.

Survival of open spaces to a period when their continued preservation has been undertaken as a municipal responsibility often owes much to the existence of terrains unfavourable for building. The alluvial flats of the Ver alongside the urban core at St. Albans are typical of many such riverine open spaces that have now been laid out as public parks. Shrewsbury, Chester, Cardiff (Figs. 5, 6, 19) and many other towns show similar features, and Princes Street Gardens in Edinburgh occupy a site that is a variant of this type. Before public consciousness of the amenity value of such tracts was sufficiently awakened or powerful to take action, however, industry and railways had laid claim to much land which, because of its physical nature, had not been encroached upon earlier. The open spaces made available by demolition of old fortifications which have sometimes been used to such advantage for loosening an over-congested urban area, as in Portsmouth (Fig. 15) or the great boulevards of some Continental cities, have also not always been immune from these encroachments.

CHAPTER VI

THE MORPHOLOGY OF TOWNS

II. THE DEVELOPMENT OF THE TOWN STRUCTURE

Planned and Unplanned Towns

IN urban studies, it is usual to emphasize as primary the difference between planned and unplanned towns. The former have been conceived and founded as towns, whereas the latter have emerged without conscious planning. They are settlements that have grown and been adapted to discharge urban functions. Their urban character has appeared in the course of their growth, and their lay-out is essentially the product of accretion of buildings about some pre-urban nucleus.

In distinguishing towns which have been created from those which have emerged, the latter have sometimes been referred to as 'spontaneous' or 'organic' towns, but the analogy between urban development and organic growth is very imperfect and should not be pressed far. It should be noted that the distinction is one which cuts across all others that derive from differences in historical development, since planned towns have appeared in many different phases as well as in many different scenes of urbanism. Among the newest recruits to the ranks of towns are examples of both types, while among the cities of the ancient world planned as well as unplanned towns were likewise represented. The Greeks regarded Hippodamus of Miletus, who lived in the fifth century B.C., as the father of town-planning, but there is clear archaeological evidence of planned cities of much earlier date, and on principles essentially similar to those applied by Hippodamus, to whom Piraeus is ascribed. Thus, whereas cities such as Ur, and the Minoan and Phoenician cities, show no sign of having been planned, Mohenjo-Daro, Khorsabad and Tel Amarna, to

name examples in three separated areas of ancient city development, were clearly planned.

The creators and designers of towns show a predilection for
certain forms of lay-out and types of construction. The latter
tend more commonly and directly to be expressions of the
architectural idiom and vogue of their period, but ideas of
civic design have often been copied from earlier masters and
may have a long history. Most widespread of all principles for
the planned lay-out of towns is the chequer-board or grid
plan, with straight parallel streets intersecting others at right
angles. Tout regarded it as natural for a planned arrangement,
but some authorities would maintain that in spite of its apparent
obviousness all applications of this plan have common ancestry.
The earliest-known representative is Mohenjo-Daro and it
was used in several other ancient cities in the Middle East.
From the sixth century B.C., when Olbia was laid out to a
grid pattern, its history is certainly continuous. It was the
standard form for the new towns established during the great
phase of Hellenistic town foundation in the Middle East
following Alexander's conquests. Adopted later by the Romans
for their colonial towns, it was spread throughout their empire.
A rectilineal pattern was not only applied to the internal
structure of the Roman town, so that it was divided into
blocks (*insulae*), the framing walls were also characteristically
laid out in the form of a square or rectangle, with four main
gates. The axial thoroughfares, known as *cardo* (north-south)
and *decumanus* (west-east), from these gates intersected at right
angles in the centre. In many towns with Roman antecedents,
even though little else may have survived, this feature is still
clearly apparent in the modern plan, e.g. Chichester, Colchester,
Tours, Orleans, Saragossa.

When planned towns reappeared with the urban revival
of the Middle Ages, the grid plan was propagated in the urban
foundations of the German colonization of the Slav countries,
where more than fifteen hundred new towns were established
during the thirteenth and fourteenth centuries. In western
Europe, too, it was often used by the founders of new towns,
notably for the 'bastide' towns of southern France, and
those established by the Christian kings in Spain as they

reconquered territory from the Moors. The grid plan is also well represented among mediaeval towns in Britain, as at Salisbury, a new town established early in the thirteenth century by Bishop Poore alongside the new cathedral, and later in Edward I's castle towns in North Wales, as well as at Hull and Winchelsea which were founded about the same time.

Renaissance and Baroque town-planning introduced new aesthetic and military principles into the designs for towns conceived in the grand manner for autocratic rulers. The grid plan did not fall into desuetude, but others, such as the spider-web or radio-centric plan, were used, as at Versailles, St. Petersburg, and Karlsruhe. It was the grid plan, however, which was adopted for the new towns established in connection with the Jacobean plantation of Ulster, and in 1682 William Penn introduced it into British North America for Philadelphia. It had been applied earlier in the Spanish colonial towns, which were laid out in rigid conformity. Nor is the grid-plan unrepresented in the East. It was used in many of the prevalent administrative cities of China, which were laid out with much ceremonial, and thence it was imported into Japan, where Kyoto exemplifies its regular system.

The open central square is a common feature of the grid-plan, and is very characteristic of the later German colonial towns. It is equally typical of the Ulster towns, where it is known as the 'diamond' and is the social hub of the town. An original English embellishment to the severity of the grid lay-out was the use of landscaped spaces to fill units of the chequer-board. These 'squares', framed by terrace-buildings, give distinction to the Georgian extensions of some of our towns.

The chequer-board or grid lay-out is so widespread a feature of the great phases of town building that it has come to be regarded as the norm for the planned town, but this association may easily be exaggerated. Moreover, while it may be represented as a formal scheme within a planned entity, as in the Roman colonial towns or the towns of the Ulster plantation, it is often simply a form of extension, filling in a framework that is irregular, and which was never conceived as of finite extent. Wherever subdivision and alienation of land are carried out on the basis of instrumental survey, rectangular lots and

D*

disposition of streets in parallel and rectilineal fashion are natural.

As opposed to planned towns, those which have evolved gradually from earlier settlements show lay-outs determined by the tenurial patterns of agricultural land, though these in turn may to a greater or less degree clearly reflect the influence of site conditions. In a few cases, the details have been studied, and their genesis has been elucidated, but in most old-established unplanned towns, the reasons for the irregularities and vagaries of the street system are lost in obscurity. At most they can only be surmised, but it is reasonable to suppose that pre-urban alignments are often perpetuated, and that if these were known, many apparently meaningless features could be explained. It has recently been pointed out how Georgetown, the capital of British Guiana, which, from its foundation at the end of the eighteenth century, has grown into a city that now approaches 100,000 inhabitants, exhibits prominently in its plan the lines of the sugar-estates which its buildings replaced. The city roads border the drainage canals or have taken their place, and are frequently punctuated by cross-streets that mark the lines of the cross-channels.

"The wards of the city correspond exactly to former plantation boundaries or are large portions of those estates sold to the city in rectangular blocks; any number of district- and street-names commemorate the old estates or their owners."

At Sydney, too, the map record which spans the growth of the settlement from its foundation in 1788 clearly shows the development of the city as a progression from the somewhat haphazard alignments established by the earliest tracks and allotments of land. In spite of his planning propensities, Governor Macquarie in the second decade of the nineteenth century found them already fixed and immutable. Thereafter the process of infilling and extension has continued grid-fashion, while preserving the essential orientation of the original lay-out and incorporating its detailed alignments, including the highly irregular nucleus at the head of Sydney Cove.

Government surveyors blocked out the environs of the original settlement of Melbourne for large pastoral holdings, and Zierer has noted that the basic lines of their first land-survey are faithfully reflected in the metropolitan street pattern. The original pastoral holdings were successively sub-divided and appropriated for building, but the principal through streets follow lines of the official survey grid. Departures from strict adherence to the straight lines and right-angles of the grid-plan are related to the irregularities of the site (the entrenchment of the valleys approaching Port Philip) and of the railways as they sought easy gradients.

The Units of the Townscape

Alike in planned and unplanned urban development, however, the scale of the units in which land has been alienated and developed for urban uses has been highly important in giving distinctiveness to the town pattern. The American town, like the Roman town, is a multiple of a standardized unit of urban area, which forms the 'block' or 'insula'. In a recent study of Hong Kong, Hughes has drawn attention to the way in which a fifteen-foot unit of street frontage dominates the pattern of Chinese towns. That it has become the common unit of land-tenure appears to be because the maximum economic length of a Chinese fir joist is about fifteen feet. This standard width for the Chinese house in town or village has often been maintained even in the reinforced concrete buildings that are now replacing the older properties. Only where Western influence is strongest, in the commercial cores of cities, is the practice departed from to any considerable extent; elsewhere

"whatever his business and however prosperous he may be the Chinese trader operates on a front of fifteen feet."

The cellular structure of an urban area, in terms of building blocks and spaces, may also be a very significant factor in its internal differentiation, as in the density zoning that is typical of residential areas of English towns. The period that followed the Public Health Act of 1875, when by-laws prescribed standards of room-size, back-space, and road-widths, is marked by considerably more open development than the preceding period, when as many houses as possible were crowded into the space available. With such improvements, the by-law housing that dominated the rapid extension of towns during the forty years before the first World War adhered to the terrace form, a debasement of its Georgian prototype, which had proved the cheapest, most convenient method of providing mass housing. Since then, the widespread vogue of Ebenezer Howard's 'garden-city' ideas has dominated suburban development, creating great tracts of open villadom, with considerable space at back and front of buildings to keep down the housing density to a prescribed figure, often standardized at twelve houses per acre. An age of motor-transport has given scope for this open form of extension which has accentuated the urban sprawl. In comparison, Continental cities have remained much more compact, but in some new countries, such as the U.S.A. and Australia, where the allowance of land for each dwelling is even more liberal, modern urbanization is still more devour-ous of land. In proportion to population, Paris is notably more compact than London, but the sprawl of London is greatly exceeded by that of Sydney and Los Angeles.

The Olden Town: the Kernel

Significant as are the differences between planned and unplanned towns, there are others no less important for urban geography which have been very much less generally recog-nized. Towns which have roots dating back to mediaeval times or beyond are notably different from those which are entirely products of the modern period, whether of overseas coloniza-tion or of the other developments that have been responsible for the modern multiplication of towns. From the standpoint of modern urban morphology a distinction perhaps even more fundamental than that between planned and unplanned towns is that between towns which originally or at some period in their history have had clearly defined extension, and those which have not. In the former, the area which represents this definitive extent is a recognizable and often fundamentally distinctive feature embodied in the present-day urban area. This kernel may or may not be a planned town, originally conceived as an entity and laid out formally, but it is hardly less distinctive if it is not.

In many towns of long standing, this kernel is the area that was formerly contained within the walls. The feature which above all else distinguished the olden town, almost if not quite wherever it occurred, was the protecting, constraining screen of its walls, through which access was restricted to a few openings where the gates were placed. Compared with the modern indefinite merging of town and country through an extensive penumbra of suburbia, the definition of the two was clear along the line of the town walls. Only in immediate proximity to the gates was any overflow typical, and there in course of time extra-mural suburbs or *faubourgs* (*fau=foris*, outside) developed. The modern town, though still needing defence, no longer finds any security in walls, which, if they do survive, have become an anachronism. The positive feature which distinguishes it is untrammelled suburban sprawl of buildings along spreading tentacles that reach into and devour the surrounding countryside.

Extension of walled towns took place in lobster-like fashion, with the periodic provision of a new and larger shell, often as much in anticipation of building extensions as to incorporate extra-mural developments that had already taken place. After some replacement of their earlier walls by extended 'enceintes' many towns had their outer limits stabilized for a long time by the very generous allowance of land enclosed within their somewhat elaborate mediaeval walls. Moreover, town populations periodically suffered drastic reductions through epidemic disease. Their very nature as concentrations of people made them especially susceptible to its ravages and their lack of sanitation made them permanent hotbeds of disease. Mortality in towns was abnormally high, and they had a notorious and merited reputation as eaters of men. Only very recently indeed have urban populations been able to grow or even be maintained by their own natural increase; in the past they have depended entirely upon constant recruitment of immigrants from the countryside. But even where developing opportunities of employment brought with them substantial increases in population there was usually space enough within the walls to accommodate the increasing numbers until the nineteenth century.

The streets of the mediaeval town might present a continuous succession of narrow buildings, forming a closed frontage, but plots of open ground extended deeply behind, and elsewhere within the town walls a number of buildings, notably those occupied by religious orders and the wealthy families, stood detached in spacious grounds. In several English towns the dissolution of the monasteries made available much urban land on the eve of a period of considerable growth of urban population. Later, the flowering of town life in Georgian Britain, when the country gentry took to the towns and established new houses there, has often left strong traces in the fabric of the town kernels. There the Georgian buildings replaced earlier structures or filled in remaining spaces of the mediaeval town, but only infrequently involved suburban extensions beyond it. On the Continent, too, the thirteenth- and fourteenth-century walling, extending beyond the limited and outgrown areas enclosed by earlier walls, was generally adequate for the growth of population until the eighteenth or even the nineteenth century.

Between the Middle Ages and the early nineteenth century, such population increase as was experienced by most towns was accommodated by crowding into the area defined by the walls rather than by expansion outside. This accounts for most of the congestion that today marks the older parts of these towns. It post-dates the mediaeval period and can give a quite misleading impression of the setting of life within a mediaeval town, where much urban land consisted of gardens, orchards and other open space.

There were, of course, exceptional towns where the mediaeval extent was exceeded long before the modern period. In some of these, as for example Vienna and Paris, continued growth was accompanied by successive circumvallations. Swelling of the kernel by the repeated process of extending the perimeter defined by walling has there provided a clearly discernible annular structure. London soon outgrew its mediaeval walls, but none of its post-mediaeval extension was enwalled. Indeed, after the definition of the city by the mediaeval wall, there was only one further extension of the bounds of the city's jurisdiction, early in the fourteenth century. This is represented by the area that comprises the wards designated by the names of the old gates with the addition of the suffix 'without'. The square mile of territory that is today the City of London includes the walled mediaeval city, together with this extension.

The prevalence and persistence of the closed town depended upon necessity. Vienna, bulwark of western Christendom against the infidel, withstood siege by the Turks as late as 1683, and long after that remained a walled city. Only a century ago, it was still confined by a double wall. The inner wall dated from the thirteenth century, and the outer, which was more or less concentric, from the early eighteenth century. The former was demolished in 1857 and replaced by the magnificent Ringstrasse, but removal of the outer wall had to wait till 1890, when it in turn was replaced by a line of boulevards, the Gürtel. Paris was provided with new walls as late as 1841 to incorporate the considerable extensions of the urban area that had taken place beyond the walls of 1785. The course of the earlier walls is marked by the Grands Boulevards, along which the Opéra and theatres were built in the nineteenth century, and the site of the 1785 wall became part of the outer boulevards. Not until after the first World War was demolition of the 1841 *enceinte* undertaken. The 200 m. military zone outside it, which had been kept free from buildings, has since provided sites for working-class housing.

The insularity of Great Britain, with the inviolable protection it provided once the internal unity of the island had been securely established, allowed an earlier liberation of towns from their walls than was usually possible on the Continent; but in Ireland, Londonderry and Coleraine were laid out as planned towns with walls early in the seventeenth century. At Londonderry, the walls remain remarkably intact to this day. In most English towns, the walling has not been reinforced since the early Middle Ages, though temporary earthen defences were in some cases thrown up hurriedly during the Civil War. Portsmouth, however, is an exception; its fortifications were thoroughly reconstructed under Charles II and remained to enclose the old town until 1876, when they gave place to the ring of barracks and parks that insulates old Portsmouth from its modern successor (Fig. 15). On the Continent the development of artillery, which ultimately was to render all walling obsolete, at first had the effect of producing increasingly elaborate and extensive systems of town fortification, such as those with which Vauban encased many towns in the eastern frontier zone of France, or the imposing forts and water defences surrounding Amsterdam and Antwerp.

The Modern Town: Urban Sprawl

The manner and scale of urban extension have undergone a remarkable transformation in modern times, largely within the past century. Previously, the growth of walled towns was essentially a process of filling in and reorganizing space within a defined perimeter, which if it required replacement by an extended girdle had the effect of producing an annular structure. This was very different from the unfettered growth, the outward sprawl, of modern towns. Ushered in by the coming of railways, and subsequently emphasized and accelerated by motor-transport on the roads, the tentacular growth of towns has proceeded at an unprecedented pace. Having in mind its geographical effects it is hardly exaggerating to liken the process to an explosion of towns into the countryside. Vast areas of landscape have been transformed into townscape; even more agricultural land, some of it of the highest quality, has been sterilized and irretrievably lost, and the arterial systems of towns have become so choked as to present some of the major physical and social problems of our age.

The census area of Greater London covers more than 700 sq. m., within which the City, mediaeval London, covers only 1 sq. m.; and even the County, created out of the adjacent counties in 1888 to include substantially the urban area which then existed, is an inner area of only 117 sq. m. The areas that constituted the whole extent of towns for centuries of their existence are often insignificant fractions of their present extent, tiny kernels embedded in a mass of enveloping tissue. Equivalent areas are now absorbed into the urban tract in a very short period. It is a stagnating town that has not grown far more in the past century than during its previous existence. In place of the small town of the past, modern developments have produced the monstrous frameless conurbation. Its advancing front of buildings engorges previously separate villages and neighbouring towns. Thereafter these once independent settlements tend simply to be sub-centres in peripheral areas of the conurbation, foci for the local shopping, entertainment, and other services of the suburbs.

The Growth and Structure of Conurbations

In their modern expansion not all conurbations have been dominated by the swelling of a single nucleus to the degree that applies, for example, in London or Paris. In the massing of people into conurbations, we may distinguish between the process of accretion, peripheral growth around a nucleus, and the process of agglomeration, coalescence of neighbouring but originally separate nuclei as they have grown and the interstices have become filled in. According to the relative prominence of one or other process conurbations show either a uninuclear or polynuclear structure. Accretion has predominated in the growth of London, although towns outside the metropolis such as Harrow, Richmond, Kingston and Croydon have been swallowed by the advancing tide of buildings. In some of our provincial conurbations, however, agglomeration has played a much more prominent role, as in the case of Stoke, the 'Five Towns' of Arnold Bennett. Administrative amalgamation here came as recently as 1908, but unification is not complete and Newcastle-under-Lyme still retains its independence of Stoke. Site conditions may powerfully assist old nuclei to retain their individuality, as in West Yorkshire (Fig. 2) where the urbanized area, heavily built up along the converging valleys of the Aire and Calder and their tributaries, straddles the watershed between in less continuous fashion, so that the conurbation as a whole lacks a central focus. Leeds, its principal member, lies towards its north-eastern margin.

The typical British conurbation represents a congeries of industrial settlements which, in the course of the past century or so, have grown and coalesced physically, but which have often retained their administrative independence. On

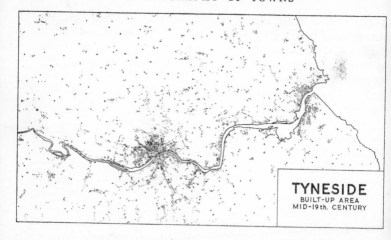

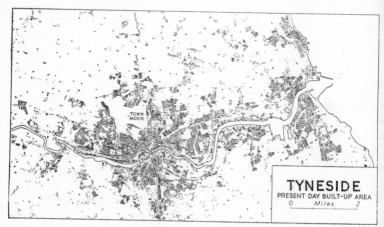

Fig. 13. The Growth of a Conurbation—Tyneside

Tyneside, for example, in the early nineteenth century there were two major concentrations of settlement, each divided by the river. The larger of these, Newcastle-Gateshead, was at the Tyne bridge, the other just within the river mouth, where North and South Shields faced each other. Between these main groups and on upper Tyneside were other industrial works and pits, with their colonies of workers, but they were individually small and separate, and green fields reached the river banks between them. Some, at least, of the nuclear cells of Tyneside had already appeared, but Tyneside was by no means yet a single conurbation. During the latter half of the century, however, expansion and multiplication brought about the merging of the riparian settlements into a continuously urbanized area, extending fourteen miles along the banks of the tidal estuary. But the administration of this conurbation is still divided among fifteen separate local authorities.

There are many examples on the coalfields of Britain, as well as in areas of intensive mineral exploitation elsewhere, e.g. the Ruhr, where settlements dependent upon individual collieries have grown and multiplied so as ultimately to present a more or less continuous mass of buildings in amorphous sprawl. In the South Wales coalfield, where pits, lines of communication, and housing are all confined to the deep-cut valleys, a definitely linear pattern has been imparted to the urbanization which now fills and even chokes the valleys. Again, along considerable stretches of the coastline, linear extension of buildings along the sea-front between once distinct resorts has fused them together. In this coastal urbanization, the older nuclei have usually kept in the forefront for shopping and entertainment facilities, and form beads in the pattern. Typically, they are the chief points where the built-up tracts along the sea-front are nourished by rail and road arteries that approach from inland.

Twin Towns and Composite Kernels

Away from the main areas of urban growth there are many scattered examples of simpler associations, usually twin towns, which duplicate some services, but share others. Several are small industrial, resort, or residential groups, which have fused without achieving concentration upon a single focus. Other twin towns are clearly distinct in primary functions and in age of development. Many of these consist of young industrial or resort settlements that have grown up within a short distance of old market towns or ports. Variant types are exemplified by Higham Ferrers and Rushden, Warwick and Leamington

Spa, Hastings and St. Leonards, Baldock and Letchworth, Reigate and Redhill, Stony Stratford and Wolverton. Even after growing together, such pairs have usually retained their administrative separateness, but whether like or unlike they inevitably count for several purposes as a single group. Of towns which are in some respects complementary, in others independent, the Medway towns present a remarkable example. To the ancient cathedral and market city of Rochester, situated on the right bank of the estuary where it is bridged by the old Canterbury-London road, there have been added the large dockyard settlements, Chatham and Gillingham. Strood is another minor element on the opposite side of the river, included in the Borough of Rochester. None of the three municipal boroughs that form the group can be said to exercise a dominant role, however, and the largest in population (Gillingham) is in most other respects a subordinate member.

The fusing of an old-established town or port with an outport nearby is an old phenomenon, exemplified by Athens-Piraeus, but it is one which has been prominent also in the development of modern conurbations, as in Edinburgh-Leith, and on a major scale in the two largest conurbations of Japan, Tokyo-Yokohama and Osaka-Kobe.

The old-established cementing of independent elements, revealed in the composite structure of the kernels of many old towns, is very different in its significance from the aggregations that result from modern urban sprawl. In the later extension of these old towns the combined elements have acted as a single nucleus, a fact not infrequently emphasized and sealed by their enclosure within a common wall. At Brunswick the wall built at the end of the seventeenth century enclosed no fewer than seven contiguous but independently walled settlements. As is well known, the heart of London has a composite structure. Near the mercantile settlement, the City of London, was the palace and abbey settlement of Westminster. Only one of these elements was ever walled, but there are numerous examples on the Continent of separately walled clusters in juxtaposition, for example, at Limoges, Aurillac, and Angers.

The structure of mediaeval society in three estates, the lay lords, clerics, and bourgeois, and the rival sponsorship of urban foundations by lay and ecclesiastical authorities are often mirrored in town-structure. But physical fusion, whether or not accompanied by complete unification, was generally accomplished early, before the phase of modern extension. Exceptionally, the union of twin towns may have been long delayed, owing to special circumstances. At Buda-Pest, for example, the river barrier perpetuated the distinction between the twin towns on opposite banks of the Danube. Buda, the administrative capital and residence of the king and nobility, and Pest, the bourgeois trading community, were each provided with walls during the thirteenth century. They experienced parallel but distinct development during the Middle Ages, and not until a bridge to connect them was built in 1849 was the way prepared for their administrative union in 1867, by which time Pest had greatly outgrown Buda.

In a previous chapter, reference has been made to the effects of the existence of a plural society upon the structure of towns. Segregation of different communities into distinct quarters with resulting duplication of services can amount to the creation of adjacent but virtually self-contained towns. Yet the interdependence of the constituent groups is usually evident in their common participation in discharging some urban activity, usually economic, which has brought them together. On the other hand, even in unified societies there are peculiar examples of twin towns, such as Minneapolis and St. Paul, existing in close proximity or even contiguity, but sharing few services and pursuing a largely separate existence.

The Shaping of Modern Extensions

To return to the simple and common case of modern extension about an old-established nucleus, the shape of the perimeter of the old town and the positions of the chief entries, especially where walls and gateways survived late, have invariably contributed to fashion the modern extension. It is important to realize that this extension has taken place along roads which were country tracks approaching the town before ever they became urban thoroughfares. Outside the Centro of Buenos Aires, the commercial core which corresponds in extent with the old colonial town of the period before 1880, the pattern of the subsequent extension has been dominated by the lines of the old pampa tracks. These provided an irregular framework for an infilling of streets laid out in regular grid fashion. A similar form of development has taken place at Sydney and Rio de Janeiro, where the devious courses of the country roads reflected their avoidance of swamp.

Beyond the built-up area at any stage of urban development the lines of communication that bind the surrounding country-side to it and provide its more distant connections are the vertebrae to which later growth has been articulated. In the process, their courses may have suffered modification, but quite often they are preserved in all their peculiarities. These, at least in part, represent adjustment to site, such as avoidance of swamp or steep gradients.

In many older towns where reconstruction of the lineaments of the pre-urban landscape in the absence of maps cannot be as well substantiated, the peculiarities of major features of the lay-out of integumental areas is doubtless to be similarly ex-plained. They are features of the pre-urban landscape which have become incorporated into the town. As established align-ments, they have played a highly important morphological role, comparable with those that directly express site conditions.

The convergences of the roads that approach a town form asters, situated on or beyond the outskirts of the old town, which have often become foci or sub-centres. Naturally, these asters often show a clear relation to the gates and the relatively early appearance of sub-foci outside the gates, associated with such features as fair-ground, smith-field, or market. Even where there never were gates, the site often imposed convergence of roads outside the town rather than within its earlier limits. At Belfast a number of sub-centres within the modern city represent the points where old tracks converged on the margin of the slobland (estuarine mud-flats), which was negotiated by a limited number of causeways. Numerous riverside towns were at a relatively early stage accompanied by a bridgehead settlement on the opposite bank of the river. This was often separated not merely by the stream itself but by an unattractive strip of accompanying flood-plain, and it provided a focus that was especially well defined. This was the relation of Southwark

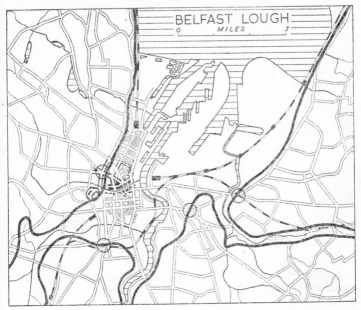

Fig. 14. Belfast

The thick black line marks the position of the old shore-line, limit of the slobland. The area already built-up by 1750 is shown in solid black. Contrast the grid pattern of the planned extensions of the early nineteenth century town on the slobland with the more irregular spread along and between the old country roads beyond its margins, where the important road asters are indicated by circle symbols

to London, and of Gateshead to Newcastle, two among countless examples of what is perhaps both the commonest and earliest type of suburb or *faubourg*.

Other new foci have been determined by the position of railway stations. Inevitably these are usually outside the old town, because of the obstacles to penetration of the kernel by railway tracks. Sometimes the site occupied by the old town precluded the railway, but elsewhere the difficulty lay simply in the compactness of the established built-up area. The railways obviously had to accommodate themselves to the geographical arrangements that existed at the time of their construction—the extent of the town and conditions of site and land-ownership in its environs as they affected opportunities for approach. Thus the margins of the built-up areas of cities at the beginning of the railway age are often indicated by the positions of the railway stations. On the north side of London, the termini are strung out from Paddington to Kings Cross along the course of the New Road which had been built in 1756 from Paddington to Islington. It marked more or less the northern front of compact building in the growing metropolis when the railways came. Open sites still left free from building on the marshy ground of the Thames flood-plain in Westminster and on the south bank gave opportunities for penetration by railways here, and provided the sites on which Victoria and Waterloo stations were built. In its earlier growth London had been largely confined to the north side of the Thames, but henceforth scope was given for the spread of south London. This took place rapidly during the second half of the nineteenth century, while failure of the railways to reach the core from the north retarded suburban extension on that side until the Underground system was provided towards the end of the century.

Many other great cities show a comparable disposition of terminal railway stations round their inner core. At Paris they form a girdle within the outermost wall of 1841. The railways even penetrated the line of the 1785 wall that is marked by the outer boulevards, but the compact urban core kept them outside the line of the Grands Boulevards which replaced the earlier walls. At Birmingham the first railways, dating from the late '30s, terminated on the eastern outskirts, the nearest they could approach without tunnelling or extensive demolition of property; but by the '50s tunnel approaches and slum clearance had brought into being the New Street and Snow Hill stations within the town centre. The earlier terminus, at Curzon Street, became a goods station.

The avoidance of valley-floors in the earlier phases of town building often provided railways both with the easy gradients they needed and with opportunities for reaching close to the hearts of towns. At Carlisle the Caldew valley brought the railways to the Citadel Station, built beside the old southern

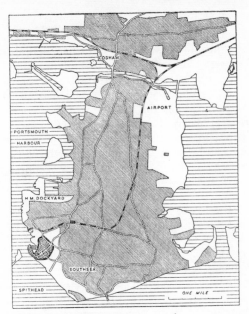

Fig. 15. Portsmouth

At Portsmouth, Commercial Rd., running north from the modern Guild-hall (G) past the railway station, has replaced the High St. in Old Portsmouth as the main axis and shopping thoroughfare

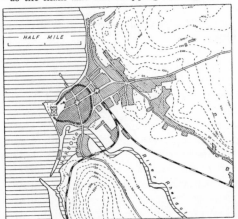

Fig. 16. Aberystwyth

At Aberystwyth the town hub (O) has shifted downhill from the middle of the "bastide" to the junction of its main exit road (Darkgate St.) with Terrace Rd., leading from railway station to sea-front

gate. At Edinburgh, between the old town, strung out on its crag-and-tail site from the Castle to Holyrood, and the new eighteenth-century town, laid out regularly on the ridge to the north, the intervening marshy valley which once contained a lake offered a course for the railway and sites for its stations. At Shrewsbury, where the old town occupies a remarkable meander-loop site, the railway approached by crossing the neck of the meander, which is occupied by the station (Fig. 5).

The railways, availing themselves of valley-lines, were followed there by the factories which depended upon them. Together they tended to create barrier belts in the residential structure of the town, separating distinct segments that came to be filled in with housing and other buildings. The approach from the old town to the new railway station often became one of the main thoroughfares of the modern town, and as such attracted shops. Considerable re-orientation of traffic and shift of retail trade within the town may have resulted. Re-centring of towns, with migration of markets and shopping localities, is not an uncommon phenomenon in their history and the railway has been one of the chief agents bringing it about. Sometimes the old town has been left as a shell from which commercial activity has largely migrated. In Portsmouth trade has left the High Street and moved to Commercial Road, which leads north past the railway station, near which the new Guildhall was built. Quite significant shifts are noticeable even in small towns. Thus at Aberystwyth the hub of the town has left the *bastide* alongside the castle headland and has moved north down Darkgate Street to the cross-roads where the main road north from the town intersects Terrace Road, leading from the railway station to the sea-front promenade.

A great many old towns which occupy cramped and inaccessible defensive sites, such as the typical perched towns of the Mediterranean countries, have in modern times been re-centred upon the railway station, or have developed in relation to the station road which connects it with the old town. Where the nineteenth-century station was placed some distance from the built-up area as it then existed, building was stimulated along and behind the station road, so that extension of the town was emphasized on the railway side as compared with

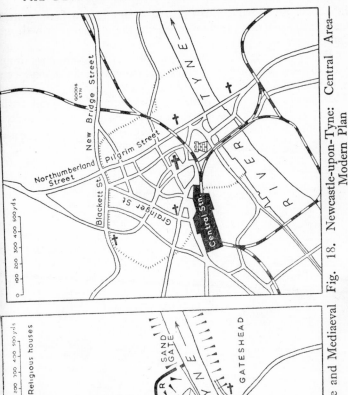

Fig. 17. Newcastle-upon-Tyne: Site and Mediaeval Plan

Fig. 18. Newcastle-upon-Tyne: Central Area—Modern Plan

the other. St. Albans illustrates this, and it is a feature that has been ever more pronounced in the modern growth of Cambridge.

Before leaving the subject of new orientations, we may refer to two examples that show a further stage in a sequence of developments related to transport changes. At Newcastle, Grainger Street developed during the course of the nineteenth century as the chief shopping street of the newly planned city. It was one of the new streets laid out by its namesake across the main thoroughfares of the mediaeval walled city. The

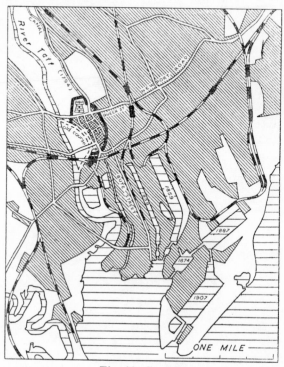

Fig. 19. Cardiff

Note the arterial system and docks in relation to the city heart in the mediaeval walled town (cross-hatching) and its modern extensions (single hatching)

atter passed from the Tyne bridge by the castle and St. Nicholas's Church to Newgate, with branches diverging to Westgate and the north gate. Significantly, Grainger Street traversed the city kernel from the Central Station in the south-west to Blackett Street, built along the line of the old north wall. Today, however, Grainger Street is rivalled or outstripped as the main shopping street by Northumberland Street. This is the northward extension, beyond the old kernel, of the arterial road which now traverses the city from the new road-bridge (1928) over the Tyne. The location of two of the three main bus stations along its line doubtless contributes to this latest internal re-orientation of the city.

At Cardiff the growth that followed the coming of the railway and the development of the town as port and service centre for the mining valleys to the north was accompanied by a shift of the main shopping thoroughfare outside the old mediaeval town to Bute Street. The latter led from its southern outskirts, where the railway station had been built, to the port created at Bute Dock. But Bute Street is no longer associated with fashionable shopping, which does not extend south of the railway. The old South Wales highroad, where it traverses the northern part of the historic town as Queen Street, has again become the main axis.

The Modification of the Kernel

The great extension of towns during the past century has been accompanied by profound modification of their older parts or kernels. Urban growth involves a process of intussusception, whereby new elements are incorporated and accommodated by replacement of existing tissue, as well as a process of accretion or agglomeration whereby additional tissue is added. Transformation of the area within the old shell and discarding of this shell are involved. But, as the example of Sydney force-fully illustrates, once a site is occupied the established arrangement is henceforth an extremely conservative influence affecting all subsequent development. Especially is this true of the street pattern. It is much easier to replace buildings than to make substantial alterations in an established ground plan. Modifications of lay-out by town-planning operations to the degree effected by Haussmann in Paris or Grainge and Dobson in Newcastle are quite exceptional. Even afte the razing of a town by pillage and fire its rebuilding is mor likely than not to follow the old street plan in all its essentials After the Great Fire of 1666, London was rebuilt not accordin to the new plan designed by Wren but in conformity with it mediaeval lay-out. After its destruction by earthquake and fir in 1923 Tokyo rose again with virtually the same street plan. I the rebuilding of these and other cities which have bee extensively blitzed in the recent war the same story is likely t be repeated. For obvious reasons that derive from legal titles t land in cities the opportunities for replanning which sucl devastation seems to offer can rarely be taken.

In modern cities the vast amount of capital sunk in provid ing the elaborate subterranean network of services add another highly conservative factor which militates agains substantial re-alignments in the lay-out above ground. Th significance of this control in city planning has recently bee emphasized by E. Randzio with special reference to post-wa Berlin (*Der Unterirdische Städtebau*, 1951). The plan of town is thus much less susceptible to change and tends there fore to be older than its profile. Thoroughfares are mor enduring than the buildings that line them and are often a old or older than the town itself. The permanence of a stree pattern is evident in the survival of Roman lines even in town where persistence of occupation through the Dark Ages i more than doubtful, and where the mediaeval town may i other respects be regarded as essentially a new creation.

On the other hand, replacement of buildings is continuall proceeding, along with alterations in the functional arrange ment. By the deliberate decision of the authorities concernec or simply by piecemeal adjustment to changed conditions seats of administration, markets and other features are liabl to move. As we have seen, even the heart or hub of the towr may suffer successive displacements in the course of time. The pattern of urban land-use as well as the actual buildings, com pared with the street pattern, are short-lived and transitory Yet the contemporary scene is always encumbered with residues, features explicable only in terms of discarded o outgrown functions and outmoded ways of life, and contributions of successive phases of the past persist even in the fabric of the present town.

Religious buildings and fortifications are often the most enduring parts of the actual fabric of old-established towns, just as they are the most conspicuous ruins of abandoned urban sites. In according special respect to buildings dedicated to religion and other sites hallowed by tradition the continual modifications and retouchings of the kernel are constrained and directed. Castles and fortifications, though substantial and durable, are less immune from neglect or deliberate destruction once their usefulness has been outlived. But every old town is likely to contain some such features of its past, either as vestigial forms or as old-established but still functioning survivals. They have exerted a powerful influence in shaping its development with each adaptation to new functions and new conditions of life, but they are important not merely for any reconstruction of the urban geography of a past phase. They have acted as architectural dominants through long periods of urban evolution. As such, they may survive in the contemporary townscape, where they often have a monumental character which imparts distinctive accidents to the urban skyline.

It is not difficult to point to towns which are mere shells exhibiting the urban forms of a bygone period rather than current manifestations of urban functions. Life lingers in them, but is not vigorous enough to be sending out new shoots or clearing away dead tissue. They are fossil towns, like Aigues Mortes. In Britain urban relics of the mediaeval cloth-trade, such as Lavenham and Long Melford in East Anglia or Burford and Chipping Campden in the Cotswolds, no longer rank as functioning towns by modern criteria of institutional equipment. That they have been able to preserve their old-world charm is essentially because of their freedom from invasion by the new urban institutions which in live towns have claimed space at the expense of older buildings. Even in vigorous, growing towns instances are not uncommon of a fossil town preserved as an enclave. Commercial functions have migrated from the old town or have developed on its outside edge. This has happened at Portsmouth and at New Orleans, where the old towns are maintained largely as museum pieces. Especially is this true of the old French and Spanish town (the Vieux Carré) embedded within the great port city of the modern American South.

Urban Unconformities

Often it is the old town, the historic kernel, which has become the commercial core of the enlarged modern urban area. Combining an old, largely original street pattern with an exceptional degree of replacement of old structures by new, the nuclear area is thus rendered especially distinctive. Its extensive modernization renders all the more striking the contrasts presented in it by surviving remnants of the past. In Buenos Aires the Centro, business core of the modern metropolis of Argentina, characterized by skyscrapers and wearing the aspect of an American city, is more or less co-extensive with the old Spanish colonial town. More often the degree of correspondence is much less exact. The limits of the old kernel have been blurred by transgression and overlap of functions. The commercial core, with its new buildings, may have spread beyond the limits of the old town. Alternatively, leaving part in the form of a residual enclave, as at St. Albans, it may include the rest of the kernel, together with an extension in some particular quarter. In either case, a new distinction is superimposed, overriding and at least to some extent effacing the pattern that expresses successive phases of urban extension. Nowhere is this more assertive than in Eastern cities, where the most striking regional contrast within the built-up area is between the integumental areas that are still typically Eastern in their adherence to traditional forms of building and the Westernized core, where these have been replaced.

Another type of urban unconformity is presented by towns which after a considerable period of stagnation and decay have been reinvigorated and launched upon a phase of new development. Such epifunctional towns, exemplified by seaside resorts that have developed from moribund ports, show a sharp juxtaposition of old and new in their structure, with an intervening period comparatively unrepresented by buildings.

The Boundaries of Urban Regions

A consideration of the nature of geographical boundaries in towns may appropriately conclude our discussion of urban regions. Functional specialization, social segregation, and historical development each provide a basis for the recognition of urban regions, and all contribute to differences in the townscape. Regional distinctiveness is most pronounced and the boundaries are most clear-cut where these patterns are co-terminous, as where the commercial core of a modern city corresponds with its historic kernel. On the other hand, what for long was a line of disjuncture along the site of old town walls, which separate urban tracts of necessarily very different age, may have been obliterated by the spread of a shopping area.

However definite the limits of an urban region may appear they are only likely to be enduring if they are reinforced by physical separation, which restricts extension and protects adjacent areas from the invasion which is so normal a feature of town development. The position of a town wall often does correspond closely with an important accident of site, such as a river-bank, a break of slope, or the junction between firm ground and low-lying floodable flats. But if it does not and the same site conditions extend beyond, so as not to present any special obstacles to extension or to suggest any differentiation of land-use, the uninterrupted spread of buildings and the functional conformity that is induced are liable to obliterate the line of disjuncture, once the walls are removed.

Regional boundaries related to site features have much more permanent significance. They persist as lineaments while the characters of the regions they separate change in the course of development. Many urban areas threaded by rivers have not always straddled them, and the historical development, land-use, and general appearance of the opposite banks usually show pronounced contrasts. Once set in train these contrasts are remarkably persistent, although the barrier effects of the rivers may in time be reduced by multiplication of bridges, as, for example, at London and Paris. Physical features, however, are themselves more subject to change by human action in towns than elsewhere. Large streams may have their courses trained and even diverted, while their tributaries are often culverted and steep-sided valleys are filled in. As with walls, although a direct influence on urban morphology is exerted only so long as the physical features persist, their indirect influence is apparent long afterwards.

Breaks in the continuity of the built-up area, however, are not all due to physical features. Other open spaces, perpetuated by old-established conditions of ownership, form enclaves and wedges which hold apart and give definition to the surrounding tracts, encouraging their separate and maybe divergent development. While natural features are not always permanent, other physical boundaries of very enduring character may be constructed in the course of a town's growth. Canals and more commonly railway tracks provide such barriers. The core of Chicago, known as the Loop, derives this name from the pattern of the elevated railway tracks which enclose it. In urban areas which have an old kernel it is of course especially in the integumental tracts developed during the nineteenth-century extension that the division by railway tracks is apparent. Unlike roads, they are in themselves considerable barriers to free movement between the areas on opposite sides. In addition, they frequently emphasize site features that tend to separate, and further insulation is often provided by the alignments of non-residential users of land, such as factories, that seek their frontages. It is not surprising, therefore, that railways should be so prominent in the delineation of the internal structure of towns. Their segregating influence is well recognized by town-planners. At Welwyn Garden City and the new town of Corby (Northants.), for example, trunk railway lines have been used to separate industrial and residential quarters. In the frequent combination of resort and residential functions in coastal towns where a railway runs near and parallel with the coast, its course often marks the separation of a coastal belt of hotels and boarding houses, together with shopping streets and places of amusement, from a more purely residential area inland.

Unlike railways which are lines of separation that naturally assume the role of regional boundaries, roads, being lines of contact, have little significance as boundaries. Main urban thoroughfares usually have frontages of similar character and are prongs of distinctive function and forms extending into otherwise homogeneous tracts; and even streets with which regional boundaries may correspond at a particular time have little permanence as such.

Only exceptionally can more than a few of the regional boundaries be either clear-cut at a particular time or stable over a considerable period. Many of the differences of character within an urban area, expressed in townscape and social structure, real and unmistakable though they may be, are of the nature of gradients. This, together with several of the features already noted, may be illustrated by reference to Middlesbrough, which has been the object of an intensive urban survey.

Middlesbrough has the distinction of being the first town created by a railway, so that its growth into a County Borough of 140,000 inhabitants is entirely a matter of the last century and a quarter, related especially to the prodigious local development of heavy industry. Lying within the great northward bend of the Tees before it opens into its estuary, the Iron-masters' District is sharply demarcated by the railway line from the rest of the town to the south. It comprises the main industrial tract, the dock and port, and the small planned town of 1830 is contained in it as St. Hilda's Ward. South of the railway the town has rapidly spread over the plain during the past century. Towards the south, the housing becomes progressively newer and socially superior. There are slums at one end, suburban villas at the other; the intermediate areas consist of buildings that are becoming obsolescent. On the east and west margins, beyond flanking shallow stream-valleys with alluvial floors, new municipal housing estates lie somewhat detached, but the main southward extension of the town over a uniform site has been remarkably compact.

The most significant hiatus in what is otherwise essentially a gradient corresponds with an interrupted belt of open spaces, including parks, cemetery, and football ground, with Albert Park (1868) in the middle. North of Albert Park, in the older part of the town, is the town centre which has developed at and near the northern end of Linthorpe Road, where railway

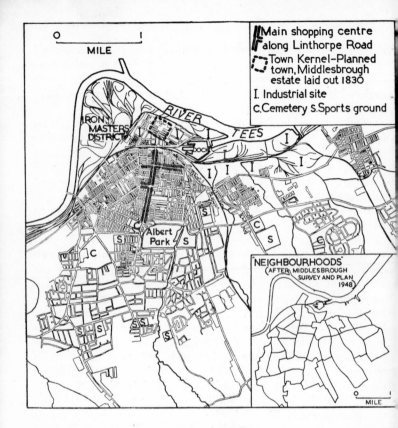

Fig. 20. Middlesbrough

The "Neighbourhoods" shown inset are urban regions, based upon homo-
geneity of physical and social characteristics, not upon social cohesion.

and bus stations, the town-hall, and the chief shops and places
of entertainment are concentrated. Not only is the hub of the
town situated here in the older area. Almost all the specifically
urban institutions—churches, clubs, public houses, restaurants,
cinemas and shops—are concentrated north of Albert Park.
The newer area to the south has some schools and most of the
playing fields, but is otherwise typical suburban housing, with
a dearth of anything else, even of shops. Compared with
24,000 houses and over 1,300 shops north of Albert Park in

1945 there were 12,000 houses but only 200 shops to the south.

The Integration of Urban Regions by Circulation

On the basis of homogeneous qualities, both of townscape and the social status of the inhabitants, there is little difficulty in recognizing urban regions in Middlesbrough, once it is accepted that in their representation on maps many, though not all, the linear boundaries shown must necessarily be somewhat arbitrary. The various characteristics that contribute to the distinctiveness of a region are not exactly co-terminous, but their patterns show a remarkable agreement. The residents of these regions, however, depend in large measure upon concentrations of work-places and central services that lie outside their bounds. Work and social life both involve much extra-regional movement. This is less pronounced and replaced by a significant degree of social cohesion only where, as in St. Hilda's Ward, a particular combination of poverty and geographical isolation reduces the mobility of the resident population.

Study of St. Albans, in many respects a very different town, showed much the same state of affairs. Although it is apparent that in terms of morphology St. Albans falls into distinct regions, with specific physical and functional characteristics (Fig. 11), there is no evidence that these are neighbourhoods in the sense of areas each equipped with its complement of primary services and possessing a measure of social cohesion. On the contrary, the essence of the internal structure of the town is geographical specialization, and the trend is towards an ever-increasing dependence upon the core for the majority of services. The urban population is highly mobile and currents of circulation criss-cross in bewildering fashion. Factories draw labour from more distant as well as from nearby housing areas, and residents seek their day-to-day shopping needs in the city centre as well as in such shops that may exist more locally. Investigation of sugar registrations under the rationing system clearly shows this to be so. And the distribution patterns of social institutions, whether public houses or churches, do not conform with the distribution of dwellings, but are striking evidence, like the relation between homes and work-places, of the amount of movement from one urban region to another. Neighbourhoods conceived as areas endowed with a degree of self-containedness, possessing within their limits the institutions used by their residents and inhabited by communities that carry out their primary social activities within the boundaries of their own territory, hardly exist. They rarely do except in the minds of town-planners. Cross-town links, dispersal of associations, and mobility are the norm—not the existence of neighbourhoods that are urban microcosms. In contrast with rural areas where the village group *is* a community by reason of its geographical isolation, the social interdependence (not independence) that characterizes the regions of towns is a product of intra-urban accessibility.

With the notable exception of towns in which rigid cleavages in the structure of society, whether of colour, language, or religion, determine the geographical segregation of distinct communities, living apart from each other, neighbourhoods scarcely exist in towns in any sense other than as urban regions with distinctive physical and functional characteristics. In the modern town-planner's sense of socially integrated areas, urban neighbourhoods require a very high degree of physical isolation. Unless they are based upon or buttressed by this or by other undesirable negative characteristics such as poverty or social segregation, it is not of the nature of urban life that they should exist. The type of human freedom which is the special gift of the urban environment is that from the bonds of the small neighbourhood group into the realization of a community life in the development of which kindred interests and personal affinities have full play. It may, therefore, be argued that it is unrealistic for planners to envisage the creation of such neighbourhoods in towns, whether existing or new. They represent a scale and type of social cohesion which, though natural to and inherent in the dispersion of groups in a rural setting, are foreign to the essentially urban quality of free association. At the same time, the need for providing suburban tracts of housing with a more adequate complement of local service institutions is undeniable.

CHAPTER VII

TOWN AND REGION
THE URBAN FIELD

Towns no longer belong to the countryside as they once did. In the pre-industrial age the great majority depended upon their immediate rural surroundings for supplies of food, and were firmly rooted in the agricultural life of the country-side. Essentially market-towns, they were not only *in* the country, but *of* it. Modern transport has removed the one-time dependence of towns upon local food-supplies, and other modern developments have brought about the multiplication on a great scale of other categories of towns, notably industrial towns and resorts, which in origin at least were quite unconnected with their rural environs.

But if the relations between town and surrounding area have been changed, they have not been severed. The two are still mutually interdependent economically and socially. Modern communications have enabled town and city greatly to extend the range of their services and have afforded the surrounding populations more direct and immediate contacts with urban life and institutions. At the same time, there has been a great increase in the number of centralized services which by their very nature find their locale in urban nodes. It is of the very essence of the modern town that it should in special degree be the seat of such centralized services, and as such it stands out in the urban-rural continuum of the present-day settlement pattern.

Townsfolk are showing an increasing tendency to settle in the country, retiring there from urban occupations or living there while actually pursuing their urban employment. Naturally, the associations of this adventitious population in the so-called 'rural' areas are essentially urban. Their very presence in villages and the countryside depends upon the opportunities which modern methods of transport offer them

to share in the employment and general amenities of town
while enjoying a more rural setting for their homes. Again, n
matter how alien by origin and outlook the industrial town c
seaside resort may be in its rural setting, many of the urba
services to which such a concentration of population gives ris
are available to a larger number of people than the civ
ratepayers and electors. They are much sought after by th
surrounding village and farm populations who find themselv
within access of shops, cinemas, and other urban services. I
fact, all towns in their essential role as service centres, n
matter what their primary *raison d'etre* as concentrations c
people, cater for an external population over and above th
town residents. Whatever the origin and occupational type c
a town, it becomes the focusing point of a district, the seat c
central services which draw to it the population of a surroundi
ing area. Towns thus exert a profound and increasing influenc
upon the economic and social groupings of the country.

Whereas a century or so ago the conscious local communit
in Britain was that of a village or a town, the development c
transport and the vastly increased number and variety c
centralized services have greatly developed the urban integra
tion of life. The range and power of the influence of towns upo
the life of the whole nation have been greatly increased. To
degree quite unknown in the past the inhabitants of the area
between the urban centres look to these towns and are draw
within their spheres of influence. They rely upon and mak
increasing use of institutions and services located there. Son
of these services are distributed from the towns to the popula
tion round about, but many must be sought, involving journey
to and from town of a frequency previously without parallel
We may go so far as to suggest that the fundamental unit in th
geographical structure of community life in a country such a
Britain is today the town region, the area whose residents loo
to a particular town as their service-centre and whose life
focused there through a constant tide of comings and going
The population living within this 'urban field', as it may l
termed for the sake of brevity, must be reckoned with the town
own inhabitants in a single community.

Indices of the Urban Field

In selecting indices by which to measure the urban field, the essential functions of urban centres must be a governing consideration, and the indices chosen must reflect these functions. Towns act as centres of employment, as collecting and marketing points for the products of the surrounding areas, and as distributing centres for goods from outside. These are their primary economic functions. Their more specifically social functions are as centres for the provision of educational, health, entertainment and cultural services; and they also provide crystallizing points of regional and district feeling and thought. As the traffic nodes of the district they are *par excellence* the meeting-places and points of assembly of the population, the hubs of its social life, and the clearing-houses of opinions and ideas.

The indices that are applicable in Great Britain will serve to illustrate the method of investigation of urban fields. Some index features that may be used are common to all towns, but others of a more specialized character are possessed only by the higher ranks of urban centre. Thus secondary grammar schools represent a centralized education service with which well-nigh every place that can lay claim to urban status in Britain is equipped, but universities are found in only a relatively few major centres. The more highly centralized institutions are symbols of high urban rank, and as such it is important to examine their service range as well as that of the more elementary urban functions. For practical purposes, moreover, it is essential that the indices should provide concrete and tangible data capable of being mapped so as to define the territorial association of town and region in respect of particular functions.

Among the minimum equipment of central services associated with townhood are secondary grammar schools and hospitals. Their service areas indicate the range of a town's functions as district educational and medical centre respectively. In the higher grades of urban centre there are large hospitals equipped with special medical services such as deep X-ray facilities, and also specialized educational institutions such as technical colleges, art schools, and in a few cases, universities. The

F

catchment areas of these will obviously provide measures of the wider regional influence of such centres.

Towns are also headquarters of numerous social organizations which are the outcome of voluntary associations. Clearly not all the social life of an urban centre is shared with the inhabitants of the surrounding district, and some associations rather than others will have a more than local significance. Among religious organizations, for instance, the denomination which shows the most strongly developed district groupings is Methodism. Like the Established Church it is far less distinctively and exclusively urban than most denominations, but whereas the social fellowship of the Church of England is primarily parochial and the basic unit of its organization quite definitely so, in Methodism the Circuit has always been of primary importance. It is a district association of churches which share a minister or ministers and a panel of lay preachers. Characteristically, it associates under the leadership of an urban centre, where the head church and residence of the superintendent minister are found, a number of outlying chapels in the surrounding district.

Another index of the field of an urban centre and one for which special importance may be claimed, is the circulation area of its newspapers.[1] In nearly all towns and indeed in many places which can scarcely claim recognition as fully developed towns, a weekly newspaper is published and distributed over the surrounding district. A large proportion of the space in such newspapers is occupied by advertisements, the great majority of which are provided by local shops, cinemas, social organizations, sales and auctions. Together with the accounts of the proceedings of the local Councils and police courts, reports of the activities of district associations, social, religious and sporting, largely make up their news contents. But besides reflecting the existence of social associations in the district, the local newspaper can itself be an important agent in promoting and focusing a sense of community. Through its medium the leadership of the town in the district is developed. Through

[1]An illuminating map and analysis of the circulation areas of local newspapers in Eire have been made by J. P. Haughton, using the local advertisements and news from villages as the main lines of evidence for determining the areas served.

he agency of its advertising, economic services located in the
own centre are made more effective throughout the circulation
area, which is thereby knit together as an economic unit.

At a considerably higher level in the urban hierarchy a daily
newspaper appears as a feature of urban equipment. More
commonly it is an evening issue, but a characteristic and outstanding feature of the provincial capital or metropolitan city
is the publication of a morning daily. Exceptional importance
attaches to daily newspapers and correspondingly to the study
of their circulation areas. They are perhaps the most potent
of all agents in the formulation and propagation of regional
opinion on important issues; and experience suggests that their
circulation areas reflect as satisfactorily as can any single
index the extreme effective range of a city's regional influence
as a community focus. This has been recognized as especially
applicable in a vast country like the U.S.A., where extensive
use has been made of newspaper circulations in defining
metropolitan regions.

It is perhaps rather surprising to find that there is so little
statistical data in Great Britain that can be applied to the
problem of determining the economic fields of urban centres.
For the information needed we are almost entirely dependent
upon personal enquiry and the co-operation of managements
of individual enterprises. Thus among all the official statistics
collected by the Census, the Board of Trade, the Ministry of
Labour and the new Ministry of National Insurance there are
none that enable us to establish the geographical relationship
between work-places and workers' homes. Any assessment of
the area a town draws upon for its labour supply depends upon
access to the employment rolls of appropriately selected concerns, and even granted the employer's goodwill and co-operation these are not always susceptible to geographical analysis.
Yet it is extremely important for planning the location of
industry and housing that there should be accurate and precise
knowledge of these relations between work-place and residence;
and the amount and extent of the influence of an urban centre as
an absorbent of labour are fundamental to a full appreciation
of its economic field.

For assessing the range of commercial functions everything

again depends upon the gauges that can be provided by representative undertakings. From information obtained from agricultural auctioneers regarding the farms from which they regularly drew livestock Dickinson was able in 1933 to map the market-areas of East Anglia. In respect of livestock market-areas there are now two different and divergent patterns. Persisting war-time controls provide us with defined fat stock market-areas, 'frozen' as they were in 1939, while the free marketing of store animals shows the rapid progress of central-ization. All fat stock within a defined Grading Area must be sent to a particular market which acts as grading-centre, but farmers have remained free to dispose of store animals in which-ever market they choose. Transport improvements, including the adoption of motor-lorry transport for the movement of livestock between farm and market, together with the enter-prise of firms of auctioneers, were rapidly effecting centraliza-tion of marketing when war broke out. Only the institution of the controlled marketing of fat stock has preserved many markets from extinction, and meantime for store animals concentration has proceeded, with elimination of redundant markets. Many of the grading-centres no longer have store sales, and the store market-areas are correspondingly larger than the Grading Areas. Thus an interesting comparative illustration is provided of the effects of the centralization which is a general trend in inter-urban relations. In Shropshire only seven markets now remain to handle the store livestock trade of the county, whereas there are fifteen grading-centres for fat stock

The outstanding and indeed the most general economic function discharged by towns is distribution. The higher ranks of urban centres are important for wholesale distribution, a centralized function in respect of which their service-areas are significant as measures of the wider field of their economic influence. But it is for shopping that the residents of a surround-ing district have most frequent, general and intimate association with the town centre. So much so that, except for daily workers 'going to town' usually means shopping. Short of a census of shoppers and their addresses, the service-area of shops can best be measured by mapping the area within which retailers undertake regular delivery of goods purchased.

The association between the shopping and entertainment functions of towns is very close. Shopping is itself a major social diversion—for women at least; but apart from this, a shopping expedition is quite commonly accompanied by patronage of some commercial entertainment. The visit to town combines business with pleasure. A town's services as market and shopping centre, as entertainment centre, and as a centre of specialized professional and social services, may all be taken advantage of on one joint-purpose visit, and are clearly interdependent. Although it may be justifiable, therefore, to regard the entertainment field of a town as more or less co-extensive with its shopping field, it is nevertheless to be regretted that it is so difficult to obtain more direct gauges of so important an urban function. The enormous development of commercial entertainment has emphasized this role of towns, since the provision of such spectacles tends to be centralized. Unfortunately, no statistics are obtained which might indicate whence the patrons of cinema, theatre, football stadium, etc., are drawn. Usually the indirect measure provided by travel facilities has to be used, making the assumption that an urban amenity that is within reach will be used. This brings us to a consideration of circulation in the urban field.

It has been implicit throughout our discussion of relations within the urban field that they involve a constant coming and going between town and region. The area thus affected by the ebb and flow of human movement is called in America the commuting area of town or city, though the term 'commuter' more especially applies to the daily traveller making his journeys to and from work. Since different transport facilities are critical for different categories of travellers, they provide clues to the importance and range of their movements. Analyses of railway season ticket and workmen's ticket issues make it possible at least in part to assess not only the extent of the area within which urban workers reside but also the volume of the movement; while issues of cheap day-return tickets give a measure of the importance of trips to town for shopping, entertainment and professional consultation by people who work and live outside the town, but who make occasional visits to avail themselves of its central services.

As bus transport has now become in many cases a much more important feature of human movements within the urban field than rail transport, it is unfortunate that bus traffic is less susceptible to statistical analysis. Considerable use, however, can be made of bus time-tables to measure accessibility, and thereby the possibility of participation by outside residents in urban facilities. Thus, buses running at various hours of the day, and on different days of the week, may have special significance for visits to town for different purposes. Late buses, leaving town at 9 p.m. or after, give some indication of the localities whose residents can take advantage of evening functions and entertainments in town; urban employment needs early morning services, earlier usually for factory workers than for office and shop workers. Shoppers can travel later, and even a daily service is not essential for shopping and marketing visits. In these ways, much can be gleaned from study of bus services. Dr. H. E. Bracey developed this type of analysis for examination of the social provision available for rural communities in Wiltshire, and other surveys have similarly used selected transport services to define the present effective range-limits for various urban functions.

Survey of Urban Fields in England and Wales

The dearth of applicable statistics and the difficulties inherent in approaching the problem partly explain why so little work has yet been done in Britain to distinguish the urban fields. We are still woefully ignorant about the areas served by various urban institutions which express the central functions of towns. The territories that are associated with urban centres for different purposes are seldom known, except vaguely. Rarely is the existing knowledge of such a precise and definite nature as to permit its being mapped, and the urban fields of Britain have never been thoroughly investigated on any extensive scale. Nor do we know much about the inter-relations of urban centres of various ranks, the degrees to which they function independently or are subsidiary to more extensively equipped centres of higher rank.

It is in an endeavour to fill this gap in our factual knowledge of the geography of Britain that a survey of urban spheres of influence is being undertaken at present under the auspices of the Geographical Association. For every settlement identified on the Ordnance Survey Quarter Inch to One Mile map of England and Wales information is being sought by use of a standardized questionnaire regarding the centres upon which the inhabitants depend for various services. The detailed questions fall into nine groups, relating respectively to education, medical, and professional services, retail distribution, cinema and other entertainment, local newspapers, agricultural markets and supplies, journeys to work and accessibility to urban centres by public transport services. Grammar schools are being used as the principal agents in collecting the information, and the detailed questionnaire forms are filled in by pupils drawn from the localities specified. The grammar schools, situated as they usually are in urban centres and drawing their pupils from catchment areas that include the surrounding settlements, provide a very satisfactory means of gathering reliable information regarding these relations. Their catchment areas are themselves one of the significant expressions of urban spheres of influence, and often provide a first approximation to the general urban field, except near county boundaries, which are limits of the areas controlled by different local education authorities. Discrepancies of extent, however, as well as exceptional towns which are not grammar-school centres or grammar-school centres with only limited development as towns, are of course revealed by the detailed answers supplied under the various headings of the questionnaire. These throw up the names of the centres to which localities are to be attributed, and at the same time show whether the association is general or exceptional.

It is hoped that the data thus being assembled will enable us to determine the nature and range of the functions of individual towns in England and Wales and throw light upon the extent to which life outside urban centres depends upon their resources. More than this, among the towns themselves the enquiry is providing evidence to what extent the equipment of services suffices the needs of the local inhabitants and to what extent they look in turn to larger towns for special features. Answers to the questions what town is commonly visited on Saturday or market-day for shopping and what larger towns, if any, are visited occasionally for special shopping prove highly significant in this respect.

While this detailed and laborious survey proceeds, Mr. F. H. W. Green has meantime carried out the mapping of what he calls the 'hinterlands' of bus-centres, as a short cut to provide town and country planners with an approximate ready indication of the spheres of influence of urban centres. A bus centre is defined as a place which has at least one regular scheduled market-day route that serves no place larger than itself, and its hinterland is defined as the area within which it is the most accessible centre. The assumptions are that within this area it will be the chief place visited by the population for central services other than those immediately available in villages; and that the picture so presented by analysis of public bus services is really representative of town-country circulation generally, and is not seriously invalidated by the existence and widespread use of private transport, such as motor-cars and bicycles. On the basis of the 1947-48 winter bus services a delimitation of average urban fields, albeit rough and ready, has thus been made for the whole country.

The Margins of Urban Fields

The shape and nature of the margins of urban fields remain to be considered. Clearly the area that can be served by or grouped with an urban centre varies greatly according to the particular service or function, so that the influence of a town over the surrounding area must have a gradient character. It does not extend with even intensity to a certain limit and then suddenly stop short. Rather are there zones of diminishing influence as the various functions are outranged. F. W. Morgan's work has demonstrated that a seaport hinterland is similarly a phenomenon of gradient and of superimposition of layers. The area served for some commodities may be very different from that for others, and plotting of the several distributions illustrates that reality is represented by a series of separate hinterlands, one for each commodity according to its nature. It is equally obvious that the limits of the service areas of towns as defined for different purposes will not coincide. The grammar school area, the hospital area, and the shopping area cannot be expected to be identical. But it is usually found that the service areas for a variety of functions correspond sufficiently closely to allow broad recognition of general or composite urban fields at a series of functional levels which accord with the more clearly defined ranks of the urban hierarchy.

Beyond range of the services discharged by all towns, the urban centre of higher ranking has an outer field or series of fields encompassing the primary fields of its smaller neighbours, for which it fulfils more specialized functions. Experience in separate plotting of the several service areas shows that the outer limits of a town's field, as defined by a variety of different criteria, present a remarkable degree of correspondence. This is far more striking than differences of detail, and clearly reflects the operation of a common factor or group of factors governing the shape and size of the urban field. Some of this correspondence of service areas is explained simply by the functions concerned being related. Thus shopping, entertainment, and professional consultation are often combined on the same visit to town, and for towns that have agricultural markets there is an obvious association of market and other central services in drawing the farming population of the surrounding district to town on market-days. But far less directly related functions, such as shopping and education, also often show very similar areas, a fact which reflects their common basis in conditions of accessibility and inter-urban competition.

In analysing accessibility it is always important to have maps to show not only the system of communications but also the public transport facilities in terms of (a) time-distance, i.e. showing isochrones, and (b) service frequency along the routes. Here we have concrete data regarding a factor which can play a decisive role in determining the range and direction of urban influence. Physical barriers can effectively isolate and put out of range tracts of territory that lie relatively near urban centres. On the other hand, arterial communications frequented by good transport services can greatly extend the availability

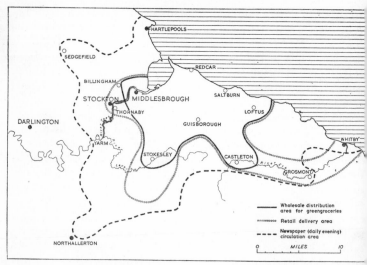

Fig. 21. The Urban Field of Middlesbrough

of urban services, along narrow strips in the case of roads or to detached outlying areas in the vicinity of stations in the case of railways. Lines of equal access to an urban centre do not form symmetrical rings, but highly irregular figures, the outermost points of which are to be found along main roads, with the possibility of detached outliers beyond. The accompanying map shows how distorted is Middlesbrough's urban field. Whereas in Cleveland to the south-east it extends effectively far along the main roads towards Whitby, and to the south along the Northallerton road beyond Stokesley, access from the nearer rural area that lies to the west, beyond the river Leven, is so hampered by the paucity of bridges that the villages there have better contacts with Stockton and Darlington than with Middlesbrough.

Absolute accessibility, however, is not the only factor of importance at work in shaping the urban field. Upon it depends rather the comparability of social provision between different parts of the territory associated with a particular centre—its gradient character. Some parts of the urban field can participate in a large range of the services provided at and from the centre; others are restricted to only a few. But not only are the limits for different functions not co-terminous; for any particular function the fields of a group of neighbouring towns often show much overlapping. The competition offered by neighbouring centres that can offer similar or better services is often decisive for the size and shape of an urban field.

In a country so closely set with towns as England, the extent of a town's sphere of influence is not usually worked out fully in all directions to the extreme limit of its several services. Instead it is cut short by entry into the field of attraction of other centres. So close is the urban mesh that before the ultimate range of one town is overstepped the pull of another is being actively exerted, so that there are marginal areas of competition between towns. The urban field is rarely developed symmetrically about a town. As an illustration, it will be seen from the accompanying map (Fig. 21) that Middlesbrough's urban field is developed very largely in Yorkshire on the south side of the river Tees. Middlesbrough's hold upon the communities of the north bank is comparatively slight and restricted. In part, this is explained by the late provision of direct communications across the river—there are no railway bridges and the two road bridges date only from 1911 and 1934. Middlesbrough may, therefore, be expected to extend and strengthen its field in county Durham. But Stockton can provide for its inhabitants and those of nearby places services hardly inferior to those of its much larger, but less well-established, neighbour. Thornaby is quite near enough to share fully many of the services provided there, but as most of these simply duplicate what Stockton offers even more conveniently, Thornaby falls within the field of Stockton rather than of Middlesbrough. No daily newspaper is published at Stockton, however, and the circulation area of the daily evening newspaper published at Middlesbrough embraces an extensive portion of south-east Durham, which for most purposes lies outside Middlesbrough's field and is served by Stockton. From an even more extensive area, patrons are drawn to first-class football matches at Middlesbrough, which is without competition between Sunderland and the West Riding in offering these spectacles.

Where a similar service is provided by two centres people living about midway between may find it convenient to use both; on the other hand, they may find both inaccessible so that neither is used. The boundaries that separate adjacent service centres are zones rather than lines, and these marginal zones are of two types: zones of overlap where duplication and alternative provision are possible, and zones of vacuum, where there is virtual absence of provision. In so far as many services must be centralized, there will always be a disparity between urban and rural standards, as indeed among urban standards. There is reduced participation in central services and progressive recession of urban influence as one service after another becomes ineffective with increasing remoteness from the urban centre. Nevertheless, it is a very significant commentary on the measure of urbanization of England and Wales that has now been attained that only about two million people do not live within five miles range of towns, as we have been using the term in the restricted sense of equipped service-centres.

Solution of the residual rural problem in the zones of vacuum that still exist hinges upon improvement of transport facilities so as to make 'urban' services available there. But few parts of England do not now enjoy easy access to towns, at least for weekly shopping and entertainment. Indeed, we can go much further than this, for an outstanding feature of our modern society is the high proportion of the nation who now live within effective range of many of the services that are available in the greatest or metropolitan cities. Metropolitan integration is far advanced in this country and few of its inhabitants live out of range of one or other of the great cities for a trip of business or pleasure that can cheaply and comfortably be accomplished within a day, allowing ample time in the city. Probably at least 90 per cent of the total population of England live within an hour's journey by public transport services of the business, shopping and entertainment centres of the city cores.

Dr. Bracey, concentrating his attention upon the functions

of towns as service-centres for rural populations as exemplified by the towns of Somerset, has devised an index of centrality based upon the affiliations of rural parishes with service centres for a selected range of shopping and professional services. Thus provided with a statistical measure for gauging the importance of rural service-centres and the orientations of individual villages in relation to them, he has been able to shed much light upon the nature of urban fields and the gradation of centres.

Around all sizeable towns there is a rather sharp passage from a core area throughout which the town is supreme and unchallenged to an outer zone within which the rural inhabitants habitually use more than one town for a considerable range of services. Beyond this again it is possible to recognize a fringe area within which some central services are used sometimes. Less important centres, however, have only small intensive or core areas, or may be without them altogether, experiencing competition from other centres right throughout their field; others again, although not without service functions are only slightly developed as centres, so that none of the surrounding parishes use them for most of the key services.

Certain towns, with highly developed district functions, such as Taunton and Yeovil in Somerset, are outstanding, with index of centrality scores far ahead of the others. The fields of the first eight towns[1] of the county, ranked on this basis, and that of Bristol together cover the whole of Somerset, but while their shopping areas frequently overlap their professional areas only infrequently do so. They are distinctive in having shopping areas that are larger than their professional areas, for they have captured much of the shopping trade of smaller country towns, but less of the professional custom. Each of these 'district centres' has a rural service-area of at least 100 square miles extent, containing an extra-urban population of at least 20,000. Their cores of intense affiliation, where little or no competition is experienced from other centres, are a substantial proportion (usually 10 per cent to 25 per cent, but in two cases more than 40 per cent) of their total

[1]Taunton, Yeovil, Bridgwater, Bath, Frome, Minehead, Wells, Weston-super-Mare.

service-areas. All these towns are visited for certain services by residents of one or more towns of lower rank. Next come other towns, 'local centres', with clearly developed service functions, as indicated both by their indices of centrality and the extent of their service areas. These local centres, however, have only small intensive areas and are further distinguished from the towns of higher grade in serving smaller districts for their shopping than for their professional services. Lastly, there are other centres, which discharge some service functions for surrounding villages, but which are clearly places of minor importance and not of urban status.

The Urban Field of Ballymena

Another example which we have investigated personally in some detail, the urban field of Ballymena in Northern Ireland, will help to emphasize some of the points already made and, at the same time, will serve to draw attention to others. Ireland is much less urbanized than England and inter-urban competition is accordingly much less strong, so that the field of a particular town is more often free to extend towards the range-limit of its services. To this fact the exceptional extent of Ballymena's urban field may be partly attributed. The town is situated in the middle of the agricultural lowlands of the county of Antrim and is a busy centre of the linen industry. Although the population resident within the confines of the municipal borough is only 14,000, the urban equipment is remarkably complete in quantity and of high quality. There are, for example, a large range of specialized shops, including representatives of several chain-stores and that hall-mark of developed urbanism in Britain, a Woolworth's store, branch offices of six banks and seven insurance companies, and a full range of professional services. A visitor to the town cannot fail to be impressed by its busy, prosperous appearance, particularly on Saturdays and market-days when it is thronged with people. Altogether both the business status and service provision are much superior to what one might expect to find in a place of its modest size.

The explanation is that, as a shopping and entertainment centre, and generally as a seat of centralized services, Ballymena

Fig. 22. The Urban Field of Ballymena

is supreme over a large area, as shown on the accompanying map. Separate mapping of the service-areas related to various functions showed a striking correspondence and revealed a territory of fairly definite extent as the general or composite urban field. Within its limits there are at least 65,000 people. Thus, for every person resident in Ballymena itself, the town serves another four people outside. The significance of this fact as compared with the population figures that are readily available may be underlined. Ballymena might appear to be among Ulster towns far inferior in general status to Londonderry, which has 50,000 inhabitants, and to be in the same class as several other Ulster towns with between 10,000 and 20,000 inhabitants, e.g. Larne, Lisburn, Lurgan and Portadown. In fact, however, the size of its urban field makes Ballymena comparable with Londonderry in urban equipment and puts it well ahead of other Ulster towns outside Belfast. Ballymena is twenty-seven miles from Belfast, far enough away to be largely independent except for the special metropolitan services. The nearest developed towns, with services at all comparable, are Ballymoney, Coleraine, Magherafelt, Ballyclare, Larne and Ballycastle, and they all lie at least sixteen miles away. The only place nearer than this which approaches urban status is Antrim, which has only 1,700 people and is provided with some, but by no means all, the characteristic features of a town.

Apart from other considerations, the coastal locations of Larne and Ballycastle limit their command of territory. It is one of those obvious facts that are sometimes overlooked that the fields of coastal towns generally must tend to be smaller than those of inland towns because so much less territory can lie within a given radius of their central services. Thanks partly to the natural advantages of its situation, but also partly to the size given it by the local concentration of industrial workers, the equipment of Ballymena, especially as regards shops and market facilities, is superior to that of most of its neighbours. In consequence we find that the territory commanded by Ballymena extends beyond the midway distance towards them. On the south-east, for example, it closely approaches Ballyclare, and on the south reaches within three miles of Antrim, even for shopping, while it includes on the coast places which are a good deal nearer Larne or Ballycastle. Ballymena competes with Larne for the allegiance of Glenarm, although the latter is more than sixteen miles away as compared with only eleven miles from Larne. The newspapers, secondary schools and technical college of Ballymena are especially extensive in their range, since they meet less competition from neighbouring towns. Only Larne and Coleraine have newspapers. Antrim, although independent of Ballymena in some respects, looks to it for both newspaper and secondary school services. From this more limited standpoint, Ballymena's urban field extends south to include the additional area which is represented on the map by hatching.

On the other hand, there are some functions which are at present restricted by limited transport facilities. Not all the potential rural clientele are effectively served by the Ballymena cinemas. As the map shows, late evening bus-services from the town do not reach all the rural area. Moreover, provision of some services is bound up with local government administration, so that their areas conform to these administrative districts. Until recently, this has been so with regard to hospital and some other health services, and it applied to the offices of the Ministry of Food, whose Control Areas adopted the framework of local government. Where the boundaries of Ballymena Rural District fall short of the town's natural sphere of influence, parts of Ballymena territory are artificially tied to other centres. This applies to an area north-west of the Long Mountain near Rasharkin, to the Antrim glens, to the area on the north shore of Lough Neagh and south of the Kells Water, and to the bridgehead areas beyond the river Bann in County Derry. The considerable bridgeless reaches of the river Bann below Lough Neagh set a definite westward limit to Ballymena's field, which coincides with the county boundary; but the bridges at Portglenone and Kilrea put the inhabitants of adjacent parts of County Derry within reach of Ballymena's services, enabling its urban field to transgress the county boundary locally.

Urban Fields and Administrative Areas

The examples just cited are typical of numerous anomalies in the administrative pattern which study of urban fields reveals. Administrative boundaries largely belong to an age before motor-transport and they ill fit the present-day facts of social geography. Before a satisfactory and comprehensive solution of the pressing problem of reform of areas in British local government can be achieved, however, accurate information about the current social and economic groupings is required.

It is apparent from what has been said about the nature of urban fields that their limits are rarely definite or stable lines. Relations between town and surrounding area are essentially fluid and we are concerned with margins that are zones, rather than lines, and which fluctuate. Urban fields do not in reality divide the country into mutually exclusive parcels like administrative units. But that does not mean that they are not relevant for the administrative structure, which ought to conform as far as is possible with the realities of community life. These include the geographical pattern, the territorial groupings that are revealed by analysis of urban fields. However carefully administrative boundaries may have been drawn to take account of real social ties—and this has not always been so—they inevitably tend to crystallize the territorial expression of community life at some particular time and are soon outdated.

While recognizing that town-country relations are always fluid, it may be pointed out, however, that major changes in the scale of the urban mesh are related to decisive stages in the development of transport, such as the coming of railways and more recently of motor-transport. The effects of the latter are still being worked out but are now sufficiently advanced for the new pattern to have been established in its essentials. It is so different from that of the administrative areas, the framework of which was set up by legislation of the late nineteenth century, as to call for drastic revision of the latter. Changed conditions of urban equipment and accessibility call for adjustments in administrative areas to remove the striking discrepancies that now exist between them and the current social regions. Further, as we have emphasized, the highly developed urban integration of life makes the urban field the real unit of modern community structure rather than the town, the village, or the rural district, whereas the local government system in its modern development is based upon a rigid dichotomy of urban and rural. The assumed antithesis of town and country has been stamped upon it, in particular by the Acts of 1888 and 1894 which imparted to it the 'island' structure which is still its essential characteristic. The larger concentrations of population have been abstracted from the old counties and set up as independent County Boroughs, and other built-up areas have likewise been set apart from their surroundings to form Urban Districts as opposed to the residual Rural Districts. The whole system divorces town from country along artificial and arbitrary lines of cleavage, inflicting upon British local government the curse of a gnawing struggle between urban and rural authorities.

It is necessary to go back to Britain before the railway age to find the contrary principle accepted. The Poor Law Unions established by the Act of 1834 applied this principle, associating rural territories with urban centres. Parishes were grouped into unions

"taking a market town as a centre and comprehending those surrounding parishes whose inhabitants are accustomed to resort to the same market".

Thus the Assistant Commissioners sent out from London to establish the Poor Law Unions reported. They chose the town

"where the medical man resided, where the Bench of Magistrates was assembling, and generally speaking the town that supplied the general wants of the district. The principal thing was convenient access to the place where the poor persons were required to attend the Board of Guardians to state their case, where the doctor lived who would supply them with medicine, and where they generally got their supplies from." (Evidence given to the Royal Sanitary Commission in 1870 by R. Weale, for thirty-three years a Poor Law Assistant Commissioner or Inspector, who had formed many of the original Unions.)

In grouping parishes thus, county boundaries were often overlapped showing that not even in the 1830s, much less today, did these boundaries correspond to natural divisions. A recent historian of local government areas has emphasized that in applying the consistent principle of deliberately grouping town and country according to ties of convenience the Poor

Law Unions present a phenomenon virtually unique in the history of English local government, but they are now essentially of historical interest. Devised before the railway age, they have been outdated by later developments that have followed the modern revolutions in transport. Indeed, many market-towns that were chosen as Poor Law Union centres in the 1830s proved inconveniently placed in the railway network. They were rapidly outpaced by neighbouring towns, and their service areas were encroached upon. Urban fields and Poor Law Unions were soon out of phase. The Unions had an average radius from market-town to boundary of about five miles; the scale of their pattern was little larger than that of mediaeval market-areas, for they still belonged to the age of unmechanized transport.

The principle upon which the Poor Law Unions were established was logical and sound; appropriately expressed in terms of contemporary institutions and social relations, it is even more valid today. Apart from the contribution an analysis and delimitation of urban fields can thus make to the problem of formulating satisfactory administrative areas, its importance to professional planners is also evident. Towns do not exist in vacuums, cut off from the contiguous areas along clear-cut municipal boundary lines. On the contrary, they are always intimately related to areas larger than the mere sites they occupy. Town and country are indivisible, both geographically and socially, and the establishment of the fundamental facts concerning their inter-relations is a condition precedent of success in the social and economic planning to which we are committed.

NOTE ON READING

The following suggestions for reading, except for a few general works, are confined to the literature available in English. The studies listed are mostly of a general character, but the opportunity is taken to include further particulars of works that are cited in the text.

GENERAL

G. Chabot, *Les Villes*, Paris, 1948.

R. E. Dickinson, *City, region and regionalism*, 1947.

* P. Geddes, *Cities in evolution*, 1915.

W. Geisler, *Die Deutsche Stadt*, Stuttgart, 1924.

P. George, *La Ville, le fait urbain à travers le monde*, Paris, 1952.

P. Lavedan, *Géographie des Villes*, Paris, 1936.

——*Histoire de l'urbanisme*, Paris, 1926–36 (4 vols.).

L. Mumford, *The Culture of Cities*, New York, 1938.

F. R. Hiorns, *Town-Building in History*, 1956.

H. M. Mayer and C. F. Kohn (editors), *Readings in Urban Geography*, Chicago, 1959.

CHAPTER I

* V. G. Childe, *The Urban Revolution*, Town Planning Review, 21, 1950.

——*What happened in History*, Pelican Books, 1942.

H. Frankfort, *Town-planning in Ancient Mesopotamia*, Town Planning Review, 21, 1950.

J. Hasebroek, *Trade and Politics in Ancient Greece* (English trans. 1933).

* H. Pirenne, *Mediaeval Cities*, 1925 (Revised English trans. 1940).

C. Stephenson, *Borough and Town: a study of urban origins in England*, 1933.

* G. T. Trewartha, *Chinese Cities*, Annals of Association of American Geographers, 41, 1951; and 42, 1952.

CHAPTER II

C. B. Fawcett, *The Balance of Urban and Rural Populations*, Geography, 15, 1929.

A. H. M. Jones, *Ancient Economic History*, 1948.

E. W. Gilbert, *The Growth of Inland and Seaside Health Resorts in England*, Scottish Geographical Mag., 55, 1939.

Chauncy D. Harris, *A functional classification of cities in the U.S.A.*, Geographical Review, 33, 1943.

United Nations Population Studies, No. 8, *Data on Urban and Rural Population in Recent Censuses*, 1950.

A. E. Smailes, *The Urban Hierarchy in England and Wales*, Geography, 29, 1944.

C. B. Fawcett, *The Distribution of Urban Population in Great Britain*, 1931, Geographical Journal, 79, 1932.

West Midland Group, *Conurbation: A survey of Birmingham and the Black Country*, 1948.

Audrey Lambert, *Millionaire Cities*, 1955, Economic Geography, 32, 1956.

F. Lorimer, *The Population of the Soviet Union*, 1946.

CHAPTER III

M. Aurousseau, *The distribution of population*, Geographical Review, 11, 1921.

A. Demangeon, *Paris, la ville et sa banlieue*, Bourrelier, Paris, 1933.

H. Ormsby, *London on the Thames*, 1924.

S. W. Wooldridge, *Some geographical aspects of the Greater London Plan*, Trans. of Inst. of British Geographers, 11, 1946.

A. G. Ogilvie, *New York and its Region*, Geography, 15, 1929.

R. N. Rudmose Brown, *Sheffield, its rise and growth*, Geography, 21, 1936.

CHAPTER IV

H. J. Fleure, *Some types of cities in temperate Europe*, Geographical Review, 9–10, 1920.

——*City Morphology in Europe*, Proc. Royal Inst., 27, 1931.

——*The Historic City in Western and Central Europe*, Bull. John Rylands Library, Manchester, 20, 1936.

H. Planitz, *Die deutsche Stadt im Mittelalter*, Graz-Köln, 1954.

G. L. Burke, *The Making of Dutch Towns*, 1956.

R. B. Hall, *The Cities of Japan*, Annals of Association of American Geographers, 24, 1934.

G. Camblin, *The Town in Ulster*, 1951.

D. Whittlesey, *Kano: A Sudanese Metropolis*, Geographical Review, 27, 1937.

——*Dakar*, Geographical Review, 31, 1941.

O. H. K. Spate, *Five Cities of the Gangetic Plain*, Geographical Review, 40, 1950.

CHAPTERS V AND VI

H. S. Thurston, *The Urban Regions of St. Albans*, Trans. of Institute of British Geographers, 19, 1953.

P. Abercrombie, *The Greater London Regional Plan*, 1944.

N. Carpenter, *The Sociology of City Life*, 1932.

Chicago Plan Commission Report, *Land Use in Chicago*, 1943.

R. E. Dickinson, *The West European City*, 1952.

C. D. Harris and E. L. Ullman, *The Nature of Cities*, Annals of American Academy of Political and Social Science, 1945.

R. E. Park, E. W. Burgess and R. D. McKenzie, *The City*, 1925.

S. E. Rasmussen, *London, the Unique City*, 1937.

——*Towns and Buildings*, 1951.

G. T. Trewartha, *Japanese Cities: Distribution and Morphology*, Geographical Review, 24, 1934.

J. K. Wright, *The Diversity of New York City*, Geographical Review, 26, 1936.

G. Taylor, *The Seven Ages of Towns*, Economic Geography, 21, 1945.

R. E. Murphy and J. E. Vance, *Central Business Districts*, Economic Geography, 30, 1954, and 31, 1955.

British Association, *Birmingham and its Regional Setting*, 1950.

M. J. Wise, *The evolution of the jewellery and gun quarters in Birmingham*, Trans. of Institute of British Geographers, 15, 1949.

M. B. Stedman, *The Townscape of Birmingham in 1956*, Trans. of Inst. of British Geographers, 25, 1958.

D. Stanislawski, *The Origin and Spread of the Grid-pattern Town*, Geographical Review, 36, 1946.

C. M. Zierer, *Melbourne as a functional centre*, Annals of Association of American Geographers, 31, 1941.

K. W. Robinson, *Sydney*, 1820–1952, Australian Geographer. 6, 1952.

G. Lighton, *Georgetown*, Geography, 35, 1950.

R. H. Hughes, *Hong Kong: an urban study*, Geographical Journal, 117, 1951.

A. E. Smailes. *Some Reflections on the Geographical Description and Analysis of Townscapes*, Trans. of Institute of British Geographers, 21, 1955.

M.R.G. Conzen, *Alnwick: A Study in Town-Plan Analysis*, Trans. of Inst. of British Geographers, 27, 1960.

Ruth Glass, *The Social Background of a Plan. A study of Middlesbrough*, 1948.

CHAPTER VII

A. E. Smailes, *The Urban Mesh of England and Wales*, Trans. of Institute of British Geographers, 11, 1946.

——*The Analysis and Delimitation of Urban Fields*, Geography, 32, 1947.

West Midland Group, *English County: A planning survey of Herefordshire*, 1946.

J. P. Haughton, *Local Newspapers and the regional geographer*, Advancement of Science, 7, 1950.

H. E. Bracey, *Social Provision in Rural Wiltshire*, 1952.

——*Towns as Rural Service-Centres—an index of centrality with special reference to Somerset*, Trans. of Inst. of British Geographers, 19, 1953.

F. H. W. Green, *Urban Hinterlands in England and Wales. An analysis of bus services*, Geographical Journal, 116, 1950.

Planning Maps (Ordnance Survey, 1/625,000), No. 6, *Local Accessibility*, with explanatory text, 1955.

F. W. Morgan, *Pre-war Hinterlands of the German Baltic Ports*, Geography, 34, 1949.

——*Pre-war Hinterlands of the German North Sea Ports*, Trans. of Inst. of British Geographers, 14, 1948.

R. D. McKenzie, *The Metropolitan Community*, New York, 1933.

V. D. Lipman, *Local Government Areas*, 1834–1945, 1949.

INDEX

A

Aberystwyth, 50, 76, 121 (Fig. 16), 122
accessibility to towns, 141–2, 145–6, 153
accretion, 27, 44, 92, 113, 125
acropolis sites, 45, 52, 122
adaptation, 86–8, 92, 95, 97
administrative areas, 36–7, 112–7, 153–6
African towns, 80, 81
agglomeration, 36, 113, 125
Aigues Mortes, 65, 127
Akkad (Agade), 11
Aleppo, 13
Alexander the Great, 16, 75, 104
Alexandria, 16, 81
American towns, 59, 60, 70, 83, 107
Amsterdam, 22, 112
Angers, 116
annular structure, 97, 110–12
Antwerp, 22, 112
Anyang, 10
Aphroditopolis, 8
Arsinoe, 17
Athens, 14, 15, 116
Aurillac, 116
Australian cities, 38
azonal factors, 97–102

B

Babylon, 10, 11, 12
Ballymena, 150–3 (Fig. 22)
barrios, 80
bastides, 76, 104
Batavia, 75
Belfast, 51, 118, 119 (Fig. 14), 152
Belgrade, 18
Berenice, 17
Berlin, 126
Besançon, 46
bidonvilles, 81
Birmingham, 93, 94 (Fig. 12), 95, 97, 120
'blocks', 107

Bombay, 53
borough, 19, 34
Bordeaux, 46
Boston, 48, 53
boulevards, 102, 111, 120
Bracey, H. E., 142, 148–9
Brasov (Kronstadt), 75
bridge-towns, 48, 56, 118
Bridgwater, 56
'bright-light district', 96
Bristol, 22, 149
Broken Hill (N.S.W.), 49
Bruges, 20, 52
Brunswick, 116
Bücher, 20
Buda-Pest, 117
Buenos Aires, 63, 75, 91, 117, 128
Burford, 127
'burgh', 19, 44
bus-centres, 144
bus-stations, 65, 125
Byblos, 12, 13
by-law housing, 108
Byzantium, 18, 82 f.

C

Cambridge, 124
canabae, 18
'cantonments', 81
capital cities, 11–12, 16, 20, 22, 23, 63
caravan cities, 13, 17
Cardiff, 82, 102, 124–5 (Fig. 19)
cardo, 104
Carlisle, 64, 120
Carrickfergus, 77, 79
Carthago Nuova, 17
castle towns, 19, 44, 73, 76, 105
cathedral cities, 19, 72, 116
Cawnpore, 59
cellular structure, 107
Chester, 45 (Fig. 6), 48, 52, 58, 102
Chicago, 56, 83, 91, 130
Chichester, 104
'Chinatowns', 82